EDUCATION IN A COMPETITIVE AND GLOBALIZING WORLD

MOBILE LEARNING

STUDENTS' PERSPECTIVES, APPLICATIONS AND CHALLENGES

EDUCATION IN A COMPETITIVE AND GLOBALIZING WORLD

Additional books in this series can be found on Nova's website under the Series tab.

Additional e-books in this series can be found on Nova's website under the eBooks tab.

EDUCATION IN A COMPETITIVE AND GLOBALIZING WORLD

MOBILE LEARNING

STUDENTS' PERSPECTIVES, APPLICATIONS AND CHALLENGES

DAVID RENÉ
AND
CORENTIN AUBIN
EDITORS

nova
science publishers
New York

Library of Congress Cataloging-in-Publication Data

ISBN: 978-1-53612-801-7

Published by Nova Science Publishers, Inc. † New York

CONTENTS

Preface **vii**

Chapter 1 Game-Based Learning to Promote Student Engagement and Motivation to Study a Novel from Classic Literature **1**
Adelina Moura

Chapter 2 Mobile Learning Scenarios in Language Teaching: Perceptions of Vocational and Professional Education Students **33**
Ana Nobre and Adelina Moura

Chapter 3 The Effects of Internet and Mobile Experiences on Mobile Learning Acceptance in Higher Education **61**
Mohamed Sarrab, Hafedh Al-Shihi, Zuhoor Al-Khanjari and Asharul Islam Khan

Chapter 4 Assessing Selection Criteria for Choosing M-Learning within a Blended Learning Course in Africa **79**
Ronald Hoyt Robertson, III

Chapter 5 Applying Mobile Technology to the Support Learning and Diagnosis Approach in a Flipped Classroom **101**
Yen-Ting Lin and Yi-Chun Lin

Chapter 6 Mobile Learning and Science Education **121**
Tiziana Maria Sirangelo

Chapter 7 Bring Your Own Device (BYOD) in Higher Education: Opportunities and Challenges **133**
Servel Miller and Katharine E. Welsh

Chapter 8 The Opportunities and Challenges of Mobile Game-Based Learning **165**
Judith Ramsay, Melody M. Terras and Elizabeth A. Boyle

Index **185**

PREFACE

Adelina Moura, PhD, begins by presenting a study wherein an educational game for Android devices was developed with the goal of motivating students to study a Portuguese author of classic literature in Chapter One. In Chapter Two, Ana Nobre, PhD and Adelina Moura, PhD review the results of two studies, one on mobile learning in teaching French as a foreign language and the other on Portuguese language as a native language. Next, Chapter Three by Mohamed Sarrab, Hafedh Al-Shihi, Zuhoor Al-Khanjari, and Asharul Islam Khan also presents a study, this time on the correlation between internet and mobile experiences and mobile learning acceptance. In Chapter Four, Ronald Hoyt Robertson III deliberates on a pilot study of the Lazy User Model to identify situations where mobile learning is more beneficial than e-Learning. Afterwards, Chapter Five by Yen-Ting Lin and Yi-Chun Lin deliberates on the implementation of mobile learning in a flipped classroom. In Chapter Six, Tiziana Maria Sirangelo presents a study on the function of mobile technologies in learning and teaching activities in science education, after which Servel Miller, PhD and Katharine E. Welsh, PhD discuss students' perceptions on Bring Your Own Device (BYOD) in Chapter Seven. Lastly, Chapter Eight by Judith Ramsay, Melody M. Terras, and Elizabeth A. Boyle examines the challenges of game-based learning.

Chapter 1 - How can schools grab students' attention and inspire them to read authors of classic literature? Will this literature still be studied in the classrooms of the future? Schools need to prepare students with real-world skills. Studying literature today is more relevant than ever. It offers students the opportunity to learn about past languages and cultures so that they better understand the present and become prepared for the uncertain future. In this context, the authors developed an educational game for Android devices to motivate students to read and study a classic author of Portuguese literature. The authors believe that, although the potential of game-based learning (GBL) remains under-valued, it can play an essential role in reshaping the students learning perception at all levels of education. In this text the authors present a learning scenario based on a serious game that was created to encourage reading among 11th year vocational course students. The game includes two complementary modes of learning: Training and Challenge. Its objective is to enable students to answer five hundred questions and fill a book of digital stickers relating to characters and locations from the plot of the novel. Finally, the authors present the students' opinions which were collected through two online questionnaires, one at the beginning and one at the end of the project. The questionnaires related to the technical and design aspects of the game and the elements of it that facilitated the study of and learning about, the novel. The results show evidence that the game was given a favourable reception and helped to increase the learning outcomes in this curricular content.

Chapter 2 - Mobile devices play a significant role in society, in general, and a very limited one at the different levels of education. Smartphones, tablets and other mobile devices allows learning to occur anywhere (and at) anytime. These powerful technological devices can enhance the teaching and learning processes by helping to promote collaborative and individual learning and broadening the boundaries of the classroom to different contexts of learning. Many students have mobile devices and their applications can provide access to learning outside the classroom, for greater flexibility and more dynamic learning. In this sense, the articulation of technological and methodological efforts allowed us to create learning scenarios supported by the devices that students take to the classroom (BYOD), and use them to motivate and involve students in meaningful learning. These devices offer the advantage of integrating various technologies in the curricular contents, such as in foreign and mother language courses, representing a set of possibilities of ubiquity that can have great impact on the learning process. Thus, the authors developed strategies with vocational and educational students' methodologies, such as augmented reality, project-based learning, game-based learning, collaborative learning and gamification. In this text, the authors present the results of two mobile learning studies in teaching French as a foreign language (to 18-23 year-old-students) and Portuguese language, as a mother tongue (to 15-19 year-old-students), in vocational education, implemented as a mediation tool in education to promote the construction of learning and development of significant skills of collaborative work. From the data collection, through a questionnaire, with open and closed questions, the authors highlight the favorable perception of the students to the integration of mobile devices in learning, and the recognition of the benefits of the teaching strategies used throughout the year, in the increase of curricular learning.

Chapter 3 - Mobile devices and applications with Internet capabilities have improved the convenience of information accessing for community of higher education learners. The learners and instructors acceptance of mobile learning is critical to the successful adoption of mobile learning therefore it is important to understand the factors that affect learners' intentions to accept mobile learning. From technology acceptance perspective, this research investigates the effect of learner's internet and mobile experiences on mobile learning acceptance in Higher education of Oman and other Arab Gulf Region. This study focused on how the level of internet and mobile experiences might influence learners' mobile learning acceptance to create a personalized learning experience in different higher education providers. The study has been conducted for 4 months from 17 different Omani higher education institutes to analyze and evaluate the relation between learners' internet and mobile experiences and mobile learning acceptance. The results of total 806 learners and instructors participated in this research showed Internet and mobile experiences have positive impact on learners and instructors' willingness and acceptance of mobile learning in different higher education providers. The effort is part of Omani-funded research project investigating the development, adoption and dissemination of mobile learning in Oman.

Chapter 4 - Mobile Learning (M-Learning) in sub-Sahara Africa has the possibility to support learners along their daily life as the implementation of smartphones in everyday activities like banking and transportation is already happening in East African countries. The here presented study used the Lazy User Model to identify in which situations in the daily life of an East African hospital worker, the usage of m-Learning provided a real benefit in comparison to using e-Learning technologies. As the Lazy User Model assumes the preferred technology of a user to access information is chosen based on the lowest amount of effort expected, the study identified reasons beyond this in the African setting. To assess this, a blended learning course was examined where learning was performed using e-Learning and m-Learning, both online and offline, whereas the course was wrapped in two face-to-face workshops. A interview guide was developed and on-site interviews have been conducted with hospital workers coming from different departments, who graduated from this course, living in Kenya and in Tanzania. The results showed that unlike the assumption of the Lazy User Model, other factors affect the opting of m-Learning over e-Learning namely security and connectivity. Factors coming from the different instructional design of these formats could not be determined in this study group. The results of this pilot study present a first assessment of identifying the most suitable format of learning materials for East African countries which play an important factor for improving both the educational sector and the provision of further training for hospital workers.

Chapter 5 - Over the last decade, mobile technology has undergone an explosion and changed educational contexts. From out-of-class to in-class, the mobile technology provides high portability and mobility to facilitate teachers and students who develop mobile learning in various educational environments. In the mobile learning environment, the mobile technology can assist teachers and students in minimizing constraints of time and place and connecting real-world and digital-world resources effectively. In recent years, to facilitate the development of students' thinking, flipped classroom is an important pedagogy that reverses in-class and out-of-class learning activities in traditional classroom. In the flipped classroom, the unidirectional delivery of teacher's instruction in the traditional classroom is changed to deploy digital learning materials on the web and students can self-learn the instruction before participating the course. Moreover, the assignment of homework and practice out of class is changed to conduct in the classroom and the teachers and students can have sufficient interactions and discussions with regard to the assignment to facilitate thinking skills during the course session. As mentioned above, it is applicable to apply mobile technology to support the flipped classroom. In practice, the mobile technology has been widely applied to support out-of-class learning in the flipped classroom. However, it is rare to concern how to apply the mobile technology to support in-class learning in the flipped classroom. In this chapter, a cross-platform learning and diagnosis web application was developed to support a software engineering course in the flipped classroom. Moreover, a study was administered to investigate the effects on students' learning motivation and performance with regard to the use of the mobile technology in the flipped classroom-based software engineering course. The investigated results revealed that applying the mobile technology to support the software engineering course in the flipped classroom that is helpful to students in enhancing learning motivation and performance. In addition, most students showed favorable perception towards the usage of the proposed application.

Chapter 6 - This study provides a critical analysis of the role that mobile technologies may play to support learning/teaching activities and in particular science education.

The opinions about their use in learning are different. Some are positive, and claim that mobile learning adapts well to the new pedagogic approaches of 'situated learning,' 'learning by doing' or 'authentic learning.' Others are more cautious and point out the almost total addiction of the youngsters to mobile technologies. Since the student uses these devices in everyday life to play and to communicate, it's unlikely he won't get distracted, preventing him from reaching concrete learning goals. Other experts insist that schools and more generally learning institutes should seize the opportunity to turn the means based on mobile technologies from distractions into learning tools. A means to do that could be integrating them in lessons and projects in the classes, providing similar learning context to the informal ones created by students in everyday life.

Developing an activity in which mobile devices are used means to elaborate learning processes where everyone, even in the collective context, may be partially protagonist. It's essential, then, to educate to a constructive use of such equipment, in which mobile technologies aren't only regarded as an entertainment or a way to communicate, but also as learning tools, a way to share, to collaborate and to support each other.

An extensive analysis of the inductive method in science education based on a student-centered approach has been carried out. It's an 'investigative' approach, in which the learner analyses the problem, observes, and formulates alternative hypothesis and comments in a critical manner. In other words, it's more like an active involvement of the student in the learning of sciences rather than a passive observation.

New technologies, including mobile ones, can't guarantee the achievement of satisfactory formative results if not coupled with well planned educational projects. However, in this context they may play a significant role and prove to be an effective support tool in several moments of sciences learning and teaching activities.

Chapter 7 - The recent growth of mobile device ownership provides significant opportunities for universities to utilise these devices to enhance the student experience. Bring Your Own Device (BYOD) allows students to use their own devices in an educational environment. This can provide opportunities such as students feeling more familiar with the technology, enabling personalisation of the devices, and could potentially reduce the cost of keeping up to date with new technology for higher education institutions. However a number of challenges also exist such as how to ensure students without mobile devices feel included and the issue of devices acting as a distraction. This chapter focuses on student's perceptions of BYOD including how motivated they feel to learn and how they use their own devices in the classroom. The findings suggest that students are willing to use their own devices, they feel more motivated to learn and they feel more connected with teaching overall.

Chapter 8 - Mobile learning has witnessed increasing adoption across a broad spectrum of educational domains. In recent years, however, a number of psychological challenges have been identified that influence the relative effectiveness of mobile learning. These include the limited nature of human information processing, including working memory capacity and the ability to attend to stimuli, especially in mobile environments that constantly change. The use of games-based learning has also increased. Whilst it is notable that in recent years the number of mobile games in particular has increased, less is known about the psychological challenges they specifically present. Recent work indicates the importance of considering the psycho-social profile of the mobile game player, the context and mechanics of the game. In this expert commentary, the authors build upon previous research into the psycho-social profile of the learner to develop a rich, nuanced understanding of the specific socio-emotional factors that influence mobile game play: these include player emotion, motivation, meta-cognition and the importance of context and how to learn, using games, across changing temporal and spatial contexts. The authors then integrate this psychological understanding with the literature on game mechanics along with the interaction between the mechanics of the game and mobile learner characteristics. The authors argue that it is through developing an insight into the intersection of mobile learning and mobile gameplay that significant gains will be made in the quest to develop effective mobile games for learning.

In: Mobile Learning
Editors: D. René and C. Aubin
ISBN: 978-1-53612-801-7

Chapter 1

GAME-BASED LEARNING TO PROMOTE STUDENT ENGAGEMENT AND MOTIVATION TO STUDY A NOVEL FROM CLASSIC LITERATURE

Adelina Moura[*]
LabTE, University of Coimbra, Portugal

ABSTRACT

How can schools grab students' attention and inspire them to read authors of classic literature? Will this literature still be studied in the classrooms of the future? Schools need to prepare students with real-world skills. Studying literature today is more relevant than ever. It offers students the opportunity to learn about past languages and cultures so that they better understand the present and become prepared for the uncertain future. In this context, we developed an educational game for Android devices to motivate students to read and study a classic author of Portuguese literature. We believe that, although the potential of game-based learning (GBL) remains under-valued, it can play an essential role in reshaping the students learning perception at all levels of education. In this text we present a learning scenario based on a serious game that was

[*] Corresponding Author E-mail: adelina8@gmail.com

created to encourage reading among 11th year vocational course students. The game includes two complementary modes of learning: Training and Challenge. Its objective is to enable students to answer five hundred questions and fill a book of digital stickers relating to characters and locations from the plot of the novel. Finally, we present the students' opinions which were collected through two online questionnaires, one at the beginning and one at the end of the project. The questionnaires related to the technical and design aspects of the game and the elements of it that facilitated the study of and learning about, the novel. The results show evidence that the game was given a favourable reception and helped to increase the learning outcomes in this curricular content.

Keywords: game-based learning, mobile devices, classic literature, motivation

INTRODUCTION

Learning is not restricted only to schools, it can be done anywhere with anyone and via different devices. Games are a form of learning, are important in physical, intellectual and social development, and are present in all cultures. Because of the evident motivational qualities of games, they have frequently been used in education as learning resources. Game-based learning dates back to the work of psychologists and pedagogues such as Piaget (1945), Vygotsky (1967), and Montessori (1974) on the role of play in child development and learning. The game is a universal activity which has been developed and applied in many different ways throughout the ages.

The most basic definition of play is an activity that has as its final goal the fun and entertainment of those who participate. No one doubts that through games one can create situations of great educational and cognitive value (Gee, 2007; Moura, 2015; Squire, 2003). Currently, the spread of computers, game consoles and mobile devices have made digital games a part of mainstream culture. With digital games we can experiment, research, solve problems, discover and reflect. Playing helps to unlock the emotional state, overcome shyness and becomes a source of motivation.

With the development of video games, the interest in their use in school learning has grown. The considerable number of publications on GBL methodology has grown (Huizenga, 2017), increasing the controversy among both supporters and opponents of the use of games in education. Part of this situation comes from the fact that there is still little knowledge of the effect of digital games on students' learning at different levels of education. The lack of empirical evidence leads to the need for more studies to help clarify the subject.

In the last decade, games have come to be used as resources for learning at different levels in different courses. Given the popularity of computer games among different levels of society, researchers have been exploring their effectiveness as learning tools. Some studies that have been conducted on electronic games show that educators and trainers intend to use gaming more in education (Fletcher & Tobias, 2012). According to these authors some results suggest that games can lead to improvements in some cognitive and psychomotor processes. For Ryu (2013), language learning through game culture, although still little studied, can expand learning opportunities while playing in an informal setting. The participants in Ryu's study revealed that, while playing they were involved in learning the language through game culture, whether through words or phrases used in the game, or through phrases and speeches used in interaction with natives or more fluent players.

Children today own and have mastered 'hi-tech' gadgets that could not have even been imagined when their parents and grandparents were children. Across the generations, work environments have modernised and today they incorporate an ever increasing amount of advanced technologies. However, there has not been a parallel incorporation of modern technologies into teaching/learning environments. So in this chapter, we present the quiz game "Os Maias: Becoming an expert!"[1] which was developed taking into account the study of the mobile game preferences of Portuguese secondary school students (Carvalho et al., 2016). The game was created for Android because, according to 2013 data

[1] "Os Maias" is the title of the Portuguese literature classic novel and the surname of the family presented in the book. We find the game in this url: http://jml.fpce.uc.pt/jogos.htm.

from the APDC (Portuguese Association for the Development of Communications), it is the most popular mobile operating system in Portugal. We hope, with the experience described in this text, to help find evidence of the benefits of mobile digital games for curricular learning purposes, such as the study of a novel from Portuguese classic literature.

Is Digital Game-Based Learning the Future of Learning?

What is meant by Game-Based Learning (GBL)? This expression is used to designate a set of works and investigations that use video games, digital games or computer games, to support learning. The gaming industry has attracted interest in the world of education, because of the playful and challenging components of games, and for their aspects that relate to social relations and acceptance within the community and specific cultures (Squire, 2003). Digital games have become increasingly complex, requiring higher thinking systems which may be of interest in formal learning.

Research in this area (Gee, 2007) has been concerned with how players learn to play such games and whether gameplay can be an effective way of organising learning activities because games offer motivating and rewarding experiences. Digital games are interactive and multimodal and may prove to be powerful learning tools, particularly games that are developed within schools. Gaming tasks are often difficult and time-consuming and the fact that children and adolescents are willing to invest considerable time and effort in performing these tasks shows that digital games should not be excluded from education. Schools should capitalise on this high level of dedication and enthusiasm and direct it towards educational ends. Games allow students to learn about themselves and the world around them. In this sense, games are considered to be important tools for the development and formation of thought, identity, values and norms.

Integrating games into education has been both encouraged and criticised over the past few decades. As regards the benefits of games for learning and cognition, gaming enthusiasts believe that players learn many important skills through digital games, and that with these skills formal learning activities will be more enjoyable, motivating and effective (Huizenga, 2017; Mitchell & Savill-Smith, 2004; Moura, 2015). The most sceptical, however, argue that computer games can have negative effects on players' attitudes and behaviours, as they often prevent people from engaging in more honourable activities such as reading, writing, or playing outdoors (Walsh et al., 2004). Despite the various concerns that are expressed about the effects of video games on children, a German study (Szycik et al., 2017) which analysed the long-term effects of the playing of video games found no connection with changes in neural responses in children. This study revealed that children do not become more aggressive or less empathic even if they play violent games every day.

The popularity of digital games has been growing, increasing sales of such games everywhere. For McGonigal (2011), digital games make us better and we can harness the power of games to solve real-world problems and make the world a better place. For this author, computer games more and more meet human needs. Through the combination of positive psychology, cognitive science and sociology, game designers have discovered the elements of games that make people happy and as a consequence they create games that take players into the virtual environments that they enjoy. As we already know, games provide rewards, stimuli, challenges and victories that are often lacking in the real world.

Since the advent of digital games, new research and publications on the subject have emerged each year showing their potential (Aldrich, 2004; Gee, 2007; Huizenga, 2017; Prensky, 2001; Squire, 2003). Research undertaken about digital game-based learning, has grown and has led to the establishment of learning principles (Gee, 2007), theories and models. But the main benefit of game-based learning has been motivation. Many games for learning have been designed to increase the motivation of the most unmotivated and difficult students. The results of a study carried out by Papastregiou in 2009 showed that the approach with games promoted the learning of curricular content and increased motivation, when compared to a group that did not use games. This study also showed that there was no significant difference between boys' and girls' performances and that games were equally motivational for both genders.

The Impact of Digital Games in Motivation

Motivation is a significant component of learning. Intrinsic motivation is an important element in education, and refers to the driving behaviour of learning, triggered by internal rewards. It arises within an individual, signifying that the learning that they are experiencing is intrinsically rewarding. This contrasts with extrinsic motivation, which leads the individual to engage in behaviour only to gain an external reward. For example, if a student reads a text because they like the subject or want to know more about a particular topic, then the student is acting on the basis of intrinsic motivation. However, if a student is reading because they have to learn about something for a class or exam or to avoid receiving a poor grade in their assessment, then their action is based on extrinsic motivation.

When do students in school have the ability to do something just for the pleasure of the activity itself? Rarely. For students, school activities are almost always obligatory requirements of either the program or the teacher. According to Kang & Tan (2014), learning that is based on virtual games can increase intrinsic motivation which in turn supports the learning process itself. The results of their study indicated that there is an increase in the impact of game-based learning on students' intrinsic motivation in relation to the subject. This is because the mental state they experience during interactive games is not limited to just the game environment, but is extended, or transferred, to the educational subject matter contained within the game. Students engage with the subject because they feel intrinsic pleasure in doing so. The results also showed that there was a significant increase in student test scores when performing gaming-based learning activities. There seems to be a direct link between increased intrinsic motivation and increased student performance.

From Video Games to Serious Games

As video games have continued to evolve, the casual games market has grown substantially. Casual gaming is a term used to refer to games that are generally free or inexpensive, only played occasionally, require minimal expertise, and are typically played on mobile devices and computers. Microsoft's Solitaire, released in 1990, is considered to be the first successful casual PC game. But what gains can digital games bring to learning?

Commercial games are rarely used in schools because teachers find it difficult to identify the importance of games to their teaching practices. A commonly stated problem regarding the introduction of games into education is that, generally speaking, educational objectives are not well-incorporated into games. In education there are goals that guide educational activities and these are very different from the objectives of games whose normal aim is to complete the tasks of the game and have fun. It is a fact that educational games are not as enjoyable as commercial games. It's also a fact that teachers do not feel prepared for using games in education. Teachers need to know how to find good games, and how to create games that are well suited to their curriculum.

For Paul Gee (2014), video games are those that are played on gaming consoles and computers. When we talk about video games we are thinking of those in which the player assumes the role of a character who moves in an elaborate virtual world, solving various problems (violent or not). Alternatively, the player creates or maintains some complex entities, such as an army, a city or a civilisation. For Gee, a good game incorporates good learning principles and is long, difficult and challenging. Squire (2011) notes that video-game learning "is about creating knowledge in a way that challenges even our most 'student-centred' notions of learning" (p. 30). Nowadays there is a growing interest in location-based games such as Pokemon Go that use GPS mobile technology.

Regarding the question of whether playing video games is a waste of time, Gee (2014) argues that when a person learns to play a video game they are learning a new literacy. While video games are popular among the younger population, these games are not designed for educational purposes. The so-called ‘serious’ or ‘educational’ games are still an insignificant part of the market and are not so popular among the players. Often, the problem is that designers have taken the fun elements out of the games, and it is therefore necessary to rethink what elements of involvement should be included in educational games (Kirriemuir & Mcfarlane, 2004). The most significant arguments in favour of the inclusion of digital games in education are that they can become a motivating and rewarding experience for students, and are a way to make learning fun, enjoyable, effective and relevant.

The serious games movement, which emerged earlier this century, was created by a group of video game enthusiasts who believe that the gaming industry has more to offer than just entertainment. Serious games are used by industries like defence, scientific exploration, health care, emergency management, city planning, engineering, education and politics. The MIT Education Arcade project[2] is a gaming project for learning. It presents a set of games for different curricular subjects, to promote deep learning and strong involvement of students in learning. These games are designed for students to learn in a safe and creative environment, to experiment, collaborate and solve problems. In addition, these games link to the curriculum and can be played on computers and mobile devices, both inside and outside the classroom.

Paul Gee (2007) has been dedicated to the study of the importance of games in the teaching and learning processes. This author presents a set of learning principles that can be identified in computer games. Given the high number of principles we will only highlight those that we consider important for education. One aspect that he stresses is that computer games put thinking and problem solving into a meaningful context. Another aspect is that the games store the thoughts and knowledge that the player acquires as they make their progress through the game. The player does not learn by moving from isolated elements to more structured and complex

[2] http://education.mit.edu/.

ones, the activity that is being performed interweaves the two. By playing, gamers learn the basic skills of the game and the game adapts itself to the player's level of experience, providing feedback which is appropriate to their level. Games can make players aware of the problems and opportunities associated with the game's strategies. Although the principles of learning mentioned by Gee may explain why the players are motivated and involved in the game's activities, it is not possible to easily transfer these principles to a school environment, since the educational objectives are totally different. But they tell us how learning environments can be organised in such a way that students understand the meanings of the various curricular disciplines.

Integrating Serious Games as Learning Resources

Serious games can be a learning tool in both formal and informal contexts. Some studies (Gee, 2007, Huizenga, 2017, Moura, 2015) show that students have a positive opinion about the use of serious games in general, and about educational games in particular. There are several types of serious games. The most interesting for the purpose of this text is the quiz-game type, such as the popular game "Who Wants to Be a Millionaire?". The attraction of quizzes lies in the fact that they are based on a latent knowledge about the subject and on the ability to find the correct answer due to the proximity or deviation of the multiple answers that are offered. The show "Who Wants to Be a Millionaire?" has increased the popularity of quizzes. Although simple quizzes do not fall into the categories of games, they can include elements of the games as in the quiz-game we present in this text. Our quiz-game features elaborate graphic design and characteristic game technology such as scores, time and opportunity to challenge other players. Other notable features of the quiz-game are the illustrations, which create an atmosphere of the past, and the opportunity for students to earn "coins", which are needed in order to enter Challenge mode.

In this world of games, there are two concepts that people often confuse. We talk about game-based learning (GBL) and gamification and they are not the same thing. GBL is learning through the use of games whereas gamification is the application of gaming techniques in non-gaming contexts with the objective of encouraging certain types of behaviour. Gamification has received great attention lately in both business and education. Although they may seem to be two similar concepts they are quite different. While games create an imaginary world in which students become immersed, gamification creates a game-like layer over real-life activities. Farber (2015) presents us with a guide to implementing GBL and gamification techniques in class and offers a set of good learning practices.

Game Presentation and Description: "Os Maias: Becoming an Expert"

The goal of any teacher is that their students both pay attention and understand school subjects. However, it is not always easy to get students' concentration. In this sense, games can be presented as learning tools that are capable of capturing the student's focus and helping them to understand the subjects of the class in a more playful and fun way. Young people are using their mobile devices more and more frequently for play. So, it makes perfect sense for schools to start taking advantage of the mobile devices, with their ever increasing performance capabilities and new features, that students bring into the classroom.

The Quiz-Game "Os Maias: Becoming an expert!" was developed for Android systems. Its target audience is 11th and 12th grade, secondary level, Portuguese course students. It can, however, be played by anybody who has an interest in the novel. The game was conceived within the scope of a funded project[3] and was in line with the curricular objectives present in the Portuguese course. The game includes all the data that's needed in order to study the novel, which was written in the second half of the 19th century by

[3] PTDC/CPE-CED/118337/2010.

the Portuguese writer Eça de Queirós. The objective of the game is for players to correctly answer the numerous (approximately 500) questions about the book, and to use their acquired knowledge to complete a booklet of digital "stickers" which relate to the novel's plot and its 45 characters and 16 map locations.

Players have to register in the game, for which they are asked to provide a username and password after which they need to select their class. Registrations only become valid after the teacher accepts them. This registration process allows teachers to access data which shows the activities that have been carried out by the students. The teacher can check the students' performance by using downloadable online Excel spreadsheets for each class. Because students are registered by class it facilitates the preparation of local and national leader boards.

The game was written in the Java programming language with the help of a game development library called "libGDX". The game has an "online" component, to record the results of each student when playing. The servers used are those of the institution owning the game[4]. The game communicates with a server through a REST (Representational State Transfer) architecture style and sends the information in JSON (JavaScript Object Notation) format.

Game Detail

After clicking the "Start" button (Figure 1)[5], the player is presented with four menu items - Training, Challenge, Digital Sticker Book and Options (Figure 2). Initially, players have to use training mode until they have earned enough coins to buy entry into challenge mode. After that they can choose to play through either training mode or challenge mode, and can access the information that they have accumulated in the sticker book and its associated map. Training mode (Figure 3) is divided into 7 'episodes' some of which are included in the syllabus and in the student handbook.

[4] http://www.fpce.uc.pt/maias/teacher/enter.

[5] All figures and tables presented in this text were created by the author specifically for this chapter.

Game Summary

The ultimate objective of any competitor is to be at the top of the leader board of their chosen challenge. In line with this idea "Os Maias: Becoming an expert!" has its own leader board and getting to the top is a matter of earning various 'rewards' by correctly answering questions about a classic novel of Portuguese literature. But there's a lot more to it than just answering questions.

The game consists of two playing modes, both of which have a 'reward' system. In 'training' mode, the rewards come in the form of coins. Access to 'challenge' mode is gained by paying an entry fee using those coins. The entry fee allows access to one of three different skill-level areas (Amateur, Experienced and Expert) where players can earn further rewards in the form of more coins, digital 'stickers' to put into a digital sticker book and, the ultimate objective, leader board points that they can boast about to their friends if they are one of the 'Top 5' players.

For each round of the game (depending on game level) 10, 15 or 25 appropriate multi-choice questions, each with four options, are randomly selected from a pool of about 500. As well as being rewarded, players can be penalised if they get too many incorrect answers. In training mode, there is no time limit, in challenge mode there is. Similar to TV's "Who Wants to Be a Millionaire", challenge mode offers players the opportunity to seek assistance if they get stuck for an answer.

Questions have been assigned degrees of difficulty which determine the size of the reward that is earned for a correct answer. However, the student is given no indication of the grade values of the questions they have answered until they have completed a round and can view their game statistics. Players can play and replay rounds from any episodes as often as they want, in order to accumulate rewards.

Figure 1. The start button.

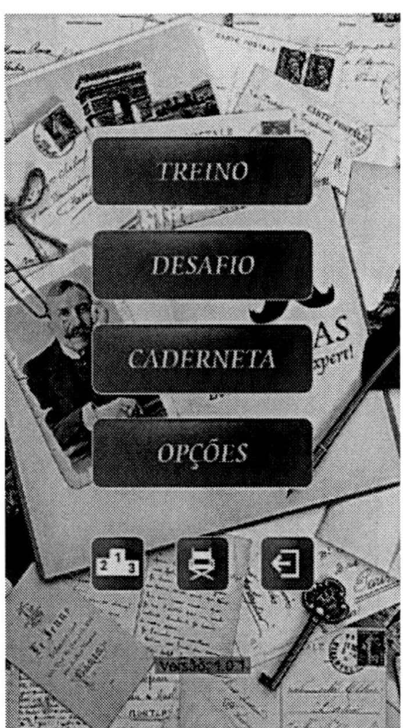

Figure 2. Initial menu.

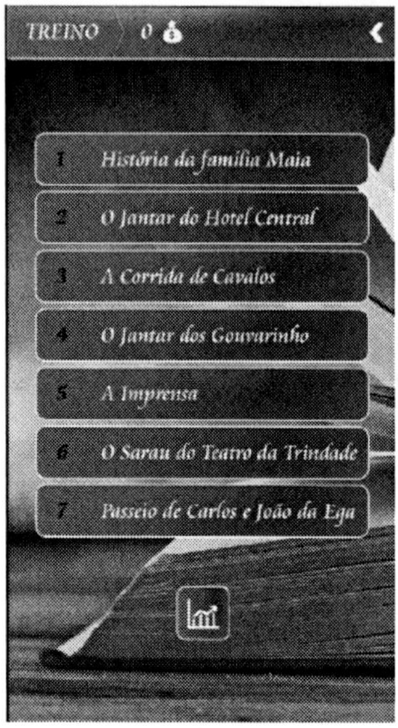

Figure 3. Training mode.

Training mode has multiple-choice questions, each with four answers, only one of which is correct. The player learns by playing for as long as they want, the goal being to earn as many coins as possible in preparation for Challenge mode. Each episode contains several rounds and each round has 10 questions. A player is allowed to get a maximum of two questions wrong. If they get a third wrong the round is terminated and they receive coins corresponding to the number of questions they have answered correctly up to that point. In order to earn more coins, players need to start a new round. In each episode the player can play whichever rounds they prefer, earning coins for each correct answer. When the questions were created, the teacher assigned one of four degrees of difficulty to each question. The more difficult the question the more coins are earned as a reward. Each time they answer a question, whether correctly or incorrectly, the player receives feedback in the form of a commentary or some text from the novel which is significant to the game (Figure 4). The objectives of providing book excerpts are to encourage the reading of the novel, and to put students in direct contact with the original text. Feedback is extremely important because it immediately guides the student and gives them useful information not only for the question itself, but also helps them to answer other questions correctly. Positive feedback for correct answers is given via a pleasant sound. For incorrect answers the player hears an unpleasant sound, alerting them to the fact that they need to concentrate harder in order to avoid further errors.

Figure 4. Feedback.

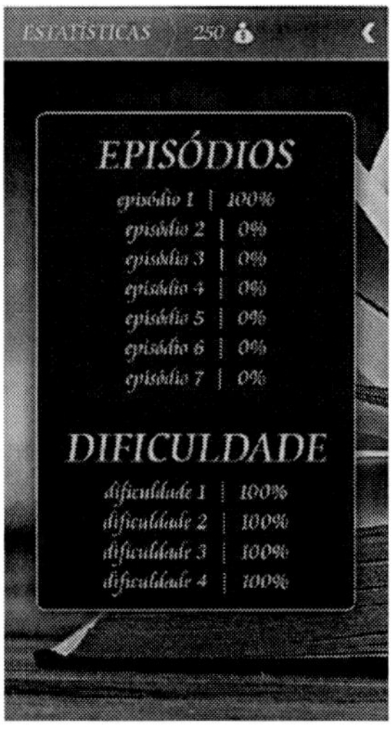

Figure 5. Statistics in training mode.

The player can interrupt the game whenever they want to and they will earn coins corresponding to the questions they have correctly answered up to that point. At the end of each round in training mode the player receives statistical information about their performance in that round (Figure 5).

In Challenge mode the player can choose between three skill levels (Figure 6):

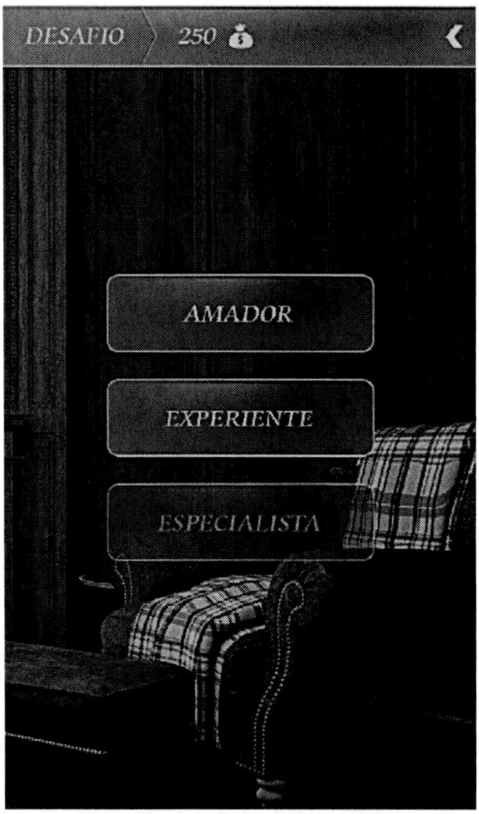

Figure 6. Challenge mode skill-levels.

Skill level	Entry Fee (coins)	Number of Questions	Question Level
Amateur	100	10	1 & 2
Experienced	300	15	1 to 4
Expert	1,000	25	3 to 6

When a player enters Challenge mode and chooses the Amateur option, they are informed that they have 1 minute to answer each question and that they can earn 20 leader board points for each correct answer (Figure 7). Figure 8 shows an example of an Amateur question with the clock having ticked down to 54 seconds. If a player gets 3 incorrect answers, they only earn the points corresponding to the last level unlocked. If they decide to stop playing, they only get the points earned up to the last correctly-answered question. If they answer all 10 questions correctly, they earn 5 stickers plus 200 coins and become able to unlock the experienced skill-level (Figure 9).

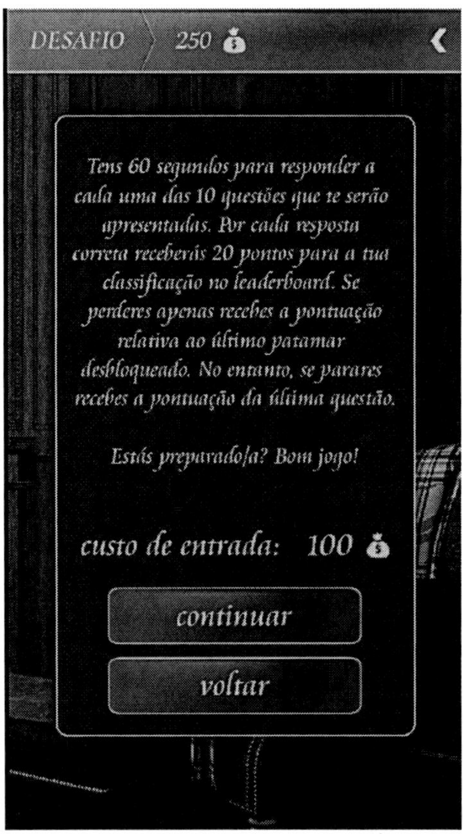

Figure 7. Amateur-level Entry fee.

Figure 8. Amateur level question.

Figure 9. Experienced level excerpt.

When playing in Challenge mode the player is not allowed to get any questions wrong, so, to compensate for this they can ask for help, as happens in the game "Who wants to be a millionaire". The player has four aids at their disposal:

- Receive an alternative question
- A poll/vote
- Consult their sticker book
- 50/50.

If the player wants to leave a round at any time, they can do so by clicking on the Exit button and they will receive the appropriate number of coins for their correctly-answered questions.

The sticker book contains the characters (Figure 10) and a map (Figure 11) which shows the locations that have been encountered by the player while in Challenge mode. Each sticker is accompanied by detailed and complementary information about the character or the map location where the plot's actions happen. There are also tips that can help the player to perform better in the Experienced skill-level in Challenge mode. Just like traditional card-trading games, the stickers for the four main characters are rarer than the others. Duplicate stickers can be exchanged for additional coins. The stickers are a form of reward for the challenges and progression of the player throughout the game.

Collecting the various characters gives players a general idea of the novel plot. The leader board (Figure 12) shows the 'TOP 5' players in that particular class. This element is favourable to the competition between players, inciting increased game practice and playing time. As they progress through the game players will accumulate points that will help them to see their name on the leader board, thereby helping them to gain the admiration of their colleagues (Social Interaction), and increase both their competitiveness and their motivation. In order to become an expert on the novel, the player can practice in Training mode, to accumulate coins and knowledge, and then play in Challenge mode to answer the maximum number of questions and thus obtain the maximum number of coins and points and hopefully collect all the stickers and finally reach 1st place in the leader board.

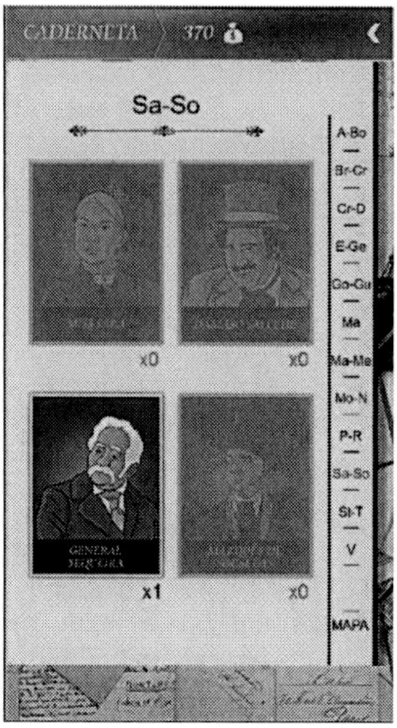

Figure 10. One activated character in the sticker book.

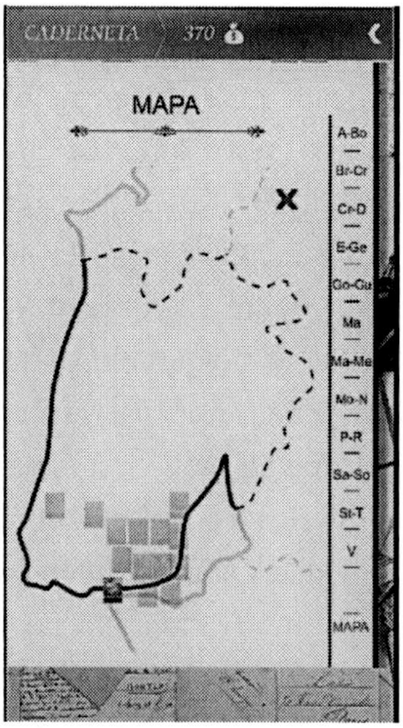

Figure 11. One activated location on the map.

Figure 12. The ‘Top 5’ leader board.

The strategies that have been designed into the game have the objectives of:

- keeping the player involved in the game (in a state of flux)
- motivating the player to progress through its skill levels (Amateur, Experienced and Expert)
- encouraging players to answer more and more questions and complete the game (by collecting the character and map location stickers).

They also allow the player to test their knowledge about the novel by offering a good balance between fun and learning, following previously-mentioned game concepts.

METHODOLOGY

Given the small number of students in the experiment, this is an exploratory, descriptive study. The evaluation approach was qualitative rather than quantitative, with the aim of capturing and analysing student's attitudes and perceptions about the game.

The evidence came from the following sources:

- a questionnaire before students had started to play the game
- a questionnaire at end of the experimental trial period
- observations made by the teacher throughout the study.

Both questionnaires were created with Google Forms and were completed online via the students' mobile devices.

Background and Participants

Thirty-five students from two classes of 11th grade vocational courses participated in the study. They were aged between 16 and 18 and comprised 6 females and 29 males. Their data were collected through the two online questionnaires mentioned above. The first survey related to the characterisation and technological profile of the participants and the second related to their opinions about the experience. These included such things as interface design, motivation, and expectations about the game. The surveys were based on Likert scale questions and open questions about learning using games. The study was carried out following earlier functionality tests, using a group of students with similar profiles to those in the actual study group, and after any necessary program alterations had been made.

Regarding the technological profile of the students:

- most of them (82%) usually access the Internet on a daily basis
- only 20% said that they surf the web for less than one hour a day

- more than 60% access the Internet both at home and at school
- the majority (60%) use the Internet to access social networks.

As regards mobile device ownership:

- all have smartphones (71% Android and 29% iOS)
- 36% also have a tablet
- 28% have their own mobile Internet connection service on their devices.

As regards game-playing habits:

- the devices used are: computer (69%), smartphone (55%) and tablet (27%)
- most students (72%) prefer to play online
- 90% usually download games to their mobile devices
- the most played games are: GTA (Grand Theft Auto), Dragon City, FIFA'17 and CS (Counter-Strike)
- 43% of the students play daily.

When the students were asked to imagine a game to study Portuguese literature, most of them said that they would like it to be a quiz-game such as "Who Wants to Be a Millionaire?". Most students (86%) considered that studying literary books from past times is important and they said that it is possible to learn school subjects by playing digital games (95%).

Analysis and Findings

In this section we discuss the results obtained from the final questionnaire, in which the students expressed their opinions about various aspects of the game, and through observations made by teacher during the two-week experimental trial period.

Students' Reaction to the Quiz-Game

The game was written for mobile devices that utilise the Android operating system and at school most students played the game using their own Android smartphones or tablets. Those students who only owned Apple's iOS-based mobile devices were loaned Android tablets for use at school. At home, they were able to play on computers by using the BlueStacks Android emulator, a piece of software that allows Android-based apps to run on computers that have either Microsoft's Windows or Apple's macOS operating systems. The overall reaction to the game was that the vast majority (95%) liked using their mobile devices to learn curricular contents.

Table 1. Technical and graphic aspects of the game (N = 35)

Item	Scale	f	%
The graphic aspect of the game ...	I really enjoyed it	17	48.6
	I liked it	18	51.4
	I did not like it	0	0
The start page of the game ...	I was very pleased	31	88.6
	I did not like it	4	11.4
The information presented at the beginning of the game was ...	Essential	21	60.0
	Important	13	37.1
	I did not notice it	1	2.9
Do you consider that the game skill levels are appropriate, in order to learn about the novel "Os Maias"?	Yes	32	91.4
	Slightly	3	8.6
	No	0	0

Table 1 shows the responses about the game's interface. The interface is very important to the success of an educational game. Thus, good navigation and usability are important to attract students' interest. All the students really enjoyed or liked the graphic aspect of the game. The game's start page pleased 88.6% of the students. The information presented at the beginning of the game (about the organisation of the game and its reward system) was essential for 60% of the students, and important for 37.1%. One student reported that they had not noticed this information. The great majority of the students (91.4%) considered that the different skill levels in the game were adequate to learn the novel.

Table 2. Motivational aspects (N = 35)

Item	Scale	f	%
When you were playing, would you see your leader board?	Yes Sometimes No	10 20 5	28.6 57.1 14.3
For you, checking the leader board was....	Stimulating Discouraging I did not go to it	28 2 5	80.0 5.7 14.3
Receiving the stickers created in you ...	The desire to continue the game The desire to have all the stickers No special interest	20 11 4	57.2 31.4 11.4
The possibility of exchanging the stickers for coins was ...	Positive Immaterial Annoying	29 6 0	82.9 17.1 0
Is earning 'coins' for correct answers a motivating element of the game?	Yes No	34 1	97.1 2.9
Did the questions, answers, stickers and coins involve you emotionally in the game?	Yes No	31 4	88.6 11.4

Regarding the motivational aspects associated with video games (Table 2), such as leader-board points, coins and stickers, 28.6% of the students frequently consulted the leader board and 57.1% sometimes consulted it. For the vast majority of students (80%) checking the leader board was stimulating, as they could track their own progress as well as that of the other top competitors. When students successfully completed a round in Challenge mode they could earn up to eight stickers representing the characters and locations that are mentioned in the story. The quantity earned depended on the number, and difficulty level, of the questions they had correctly answered. For most of the students (57.2%) the stickers made them want to continue playing and 31.4% wanted to get all the stickers. The ability to exchange stickers for coins was a positive feature for most of the students (82.9%) and immaterial to the others. For almost all students (97.1%) earning coins for the right answers was a motivating element of the game. The vast majority of students (88.6%) considered that the questions, answers, stickers and coins made them feel emotionally involved with the game.

Table 3. Curricular learning (N = 35)

Item	Scale	f	%
With the game "Os Maias" ...	I learned a lot I learned little I learned nothing	34 1 0	97.1 2.9 0
Was playing the game worthwhile?	Yes No	35 0	100.0 0
Did you consider the information given in the comments at the end of each question to be important?	Yes No	35 0	100.0 0
Did you feel that the explanatory texts that accompanied the stickers helped you to learn?	Yes Slightly No	27 8 0	77.1 22.9 0
Has playing the game "Os Maias" motivated you to look for more information?	Yes No	28 7	80.0 20.0
Has playing the game made you curious to read the novel?	Yes Slightly No	25 6 4	72.0 17.0 11.0
Did the game's positive and negative feedback help you to learn?	Yes No	32 3	91.4 8.6
Learning about the novel "Os Maias" via a game …	Was fun Helped me learn a lot in a short time. I did not like it	28 7 0	80.0 20.0 0
Will you recommend the game to your friends, for them to learn more about the novel?	Yes Perhaps No	26 8 1	74.0 23.0 3.0

Table 4. Students' Open Comments

Student 1	The game is good for learning about this novel.
Student 2	I really liked it, we should have more games like that.
Student 3	This way of learning is very useful. It attracts us because we are living in the age of technology. I loved playing it, I highly recommend it.
Student 4	I learned a lot with this game, it's the most fun way to learn.
Student 5	I liked everything in this game because it helped me to study.
Student 6	The game is a lot of fun and a simple and effective way to learn about this novel.
Student 7	I just did not like it when I was wrong.
Student 8	I really liked the game, it's simple to use and much fun.

Our observations revealed that the students helped each other when they played in training mode. The more advanced students helped their weaker colleagues to get the right answers and thus earn more coins. In all students, this interaction generated both enthusiasm and the motivation to play and learn (social learning).

Table 3 shows the students' opinions about curricular learning that this game generated. Almost all the students (97.1%) said they learned a lot about the literary novel. All the students said that playing "Os Maias" was worthwhile. They thought that the information which appeared in the comments at the end of each question (whether or not it was answered correctly) was important. This feedback was considered to be an enrichment of learning. Regarding the information that accompanied each sticker, the majority of the students (77.1%) thought that it helped them to learn, the remaining students said that it helped them slightly with their learning. Most of the students (80%) said that playing the game motivated them to look for more information about the book. As regards whether playing the game stimulated their curiosity to read the original book, 72% said yes, 17% said slightly and 11% said no. These data are important because one of the goals of the game was to motivate students to read the full book. The vast majority of students (91.4%) reported that feedback, positive or negative, helped them to learn the novel. As most of the feedback comments were excerpts from the original book itself, or illustrative clarifications, they helped to reinforce the information that was contained in each question. For the majority of the students (80.0%) learning Eça de Queirós' novel through the "Os Maias" game was fun and for 20.0% of them it allowed them to learn a lot in a short time. This shows the effectiveness of this particular game in learning this particular curricular subject. The majority of students (74%) would recommend to friends that they should play the game to learn more about this content.

As can be seen in Table 4, the students thought that playing this game was an easy and fun way to learn a literary novel. Almost all students expressed a positive opinion in the comments on this learning experience and on the game itself. This is confirmed by the observations we made during the classes in which the study was developed. These positive perceptions of the game are in line with the study done by Moura (2015), using Arcade games in mother tongue and foreign language classes. This encourages us to use more games in the future to help improve students' learning in curricular subjects such as the study of a classic novel.

At the end of the experiment the students took a quiz-test with 50 questions about the book, similar to the questions that appeared in the game. Most students were able to get good results. However, the students who spent the most time playing and who were more interested obtained the better scores.

CONCLUSION

Most students come to school with an experience of many hours spent playing online video games with players from around the world. They would like it if school courses offered them the opportunity to also have gaming experiences as part of their learning. Quiz-games have previously been shown to be resources that can enhance students' learning experiences. The main objective of this study was to evaluate the effective learning and motivational effects of an Android quiz-game about a classic 19th century Portuguese literary novel. The participants in the study were students from two 11th grade vocational course classes. They were observed throughout the time that they played the game in the classroom. The opinions of the students were collected by two online questionnaires, one before they encountered the game and one after they had finished playing the game both inside and outside the classroom. With this game we intended to bring the students closer to the study of this literary novel, and to make the game relevant to formal learning tasks. The focus of the study was not so much on the game itself but on its usefulness as a resource that teachers can embed in learning activities such as the study of a literary novel.

The results suggest that serious games can be exploited as effective and motivational learning environments. From our observations, we can verify that the game had the same motivational impact on both genders as were indicated in Papastergiou's results in 2009. We believe that this game helped to facilitate the learning of a curricular content that is usually considered difficult and tedious by the students, particularly those students who find it challenging to read the works of classic authors. We believe that the students learned more because they became involved in the game when it was installed on their mobile devices. The results show that GBL helped to increase the students' level of involvement, participation and collaboration in the accomplishment of school activities and peer learning. We also found that the students exhibited superior levels of commitment, and involvement in the study of the book, compared to that which is found in more traditional classes.

Although limited, the study shows that it is necessary to continue investigating this subject in order to gather more robust empirical evidence on the advantages and disadvantages of using educational games in school. The future will be different, because jobs prevalent today will be obsolete tomorrow. Schools must transform and follow the trends that exist in the society of which they are a part. Teachers must be prepared to incorporate games into their practices.

REFERENCES

Aldrich, C. (2004). *Simulations and the future of learning: An innovative (and perhaps revolutionary) approach to e-learning*. San Francisco: Pfeiffer. Bright.

Carvalho, A. A., Cruz, S., Barros, C., Moura, A., Araújo, I. & Zagalo, N. (2016). Aspetos a considerar na criação de jogos educativos. In Carvalho, A. A. A.; Cruz, S.; Marques, C. G.; Moura, A.; Santos, M. I., & Zagalo, N. (2016) (orgs). *Atas do 3º Encontro sobre Jogos e Mobile Learning*. Coimbra: Universidade de Coimbra, FPCE, LabTE, pp. 510-518. [Carvalho, A.A, Cruz, S., Barros, C., Moura, A., Araújo, I., Zagalo, N. (2016). Aspects to Consider in Creating Educational Games. In Carvalho, A.A.A.; Cruz, S.; Marques, C. G.; Moura, A.; Santos, M. I., & Zagalo, N. (2016) (orgs), *Proceedings of the 3rd*

Conference on Games and Mobile Learning. Coimbra: University of Coimbra FPCE, LabTE, pp. 510-518].

Fall, Anna-Mária. & Roberts, G. (2012). High school dropouts: Interactions between social context, self-perceptions, school engagement, and student dropout. *Journal of Adolescence*, Volume *35*, Issue 4, pp. 787-798. Available from: https:// pdfs.semanticscholar. org/24c1/ e51de77e 47591ef71e7978223aa432989cb5.pdf.

Farber, M. (2015). *Gamify your classroom: A field guide to game-based learning*. New York: Peter Lang Publishing, Inc.

Ford, V. B. & Roby, Douglas E. (2013). Why Do High School Students Lack Motivation in the Classroom? *Global Education Journal*, Vol. 2013 Issue, *2*, p. 101. Available from: https://libres.uncg.edu/ir/uncp/f/ Why%20Do%20High%20School%20Students%20Lack%20Motivatio n%20in%20the%20Classroom.pdf.

Gee, J. P. (2007). *Good Video Games + Good Learning: Collected essays on video games, learning and literacy*. New York: Peter Lang.

Gee, J. P. (2014). *What video games have to teach us about learning and literacy*. New York: Palgrave Macmillan.

Huizenga, J. C. (2017). *Digital game-based learning in secondary education*. Available from: http://dare.uva.nl/search? identifier= c2bf59b2-82f0-4791-ae50-981865e955aa.

Kang, B. & Tan, S. H. (2014). Interactive games: intrinsic and extrinsic motivation, achievement, and satisfaction. *Journal of management and strategy: JMS*, Vol. *5*, nº 4, pp. 110-116.

Kirriemuir, J. & McFarlane, A. (2004). *Literature Review in Games and Learning*. Bristol: Nesta Futurelab series, report, *8*.

Legault, L., Green-Demers, I. & Pelletier, L. (2006). Why Do High School Students Lack Motivation in the Classroom? Toward an Understanding of Academic Amotivation and the Role of Social Support. *Journal of Educational Psychology*, 2006, Vol. *98*, No. 3, 567–582. Available from: http:// selfdeterminationtheory.org/SDT/ documents/ 2006_ LegaultGreenPelletier_JEP.pdf.

McGonigal, J. (2011). *Reality is Broken: Why Games Make Us Better and How They Can Change the World*. New York: Penguin Press.

Mitchell, A. & Saville-Smith, C. (2004). *The use of computer and video games for learning. A review of the literature*. London: The Learning and Skills Development Agency.

Montessori, M. (1974). *Childhood Education*. Chicago: Henry Regnery.

Moura, A. (2015). Using arcade games to engage students in the learning of foreign and mother languages. In Giovanni Vincenti & James Braman (Eds.), *EAI Endorsed Transactions on e-Learning*, (Vol. *2*), Issue 5, e2.

Piaget, J. (1945). *Play, dreams and imitation in childhood*. London: Heinemann.

Prensky, M. (2001). *From Digital Game-Based Learning*. McGraw-Hill.

Ryu, D. (2013) Play to learn, learn to play: Language learning through gaming culture. *The Journal of Eurocall*, *25*(2), 286–301.

Squire, K. (2003). Video games and education. *International Journal of Intelligent Simulations and Gaming*, *2* (1), 49-62.

Squire, K. (2011). *Video games and learning: Teaching and participatory culture in the digital age*. New York: Teachers College Press.

Szycik, G. R., Mohammadi, B., Münte, T. F. & te Wildt, B. T. (2017). Lack of Evidence That Neural Empathic Responses Are Blunted in Excessive Users of Violent Video Games: An fMRI Study. *Journal Front. Psychol., 08*.

Tobias, S. & Fletcher, D. (2012). Learning from Computer Games: A Research Review. In De Wannemacker S., Vandercruysse S., Clarebout G. (eds), *Serious Games: The Challenge. Communications in Computer and Information Science*, vol. 280, pp-6-17. Springer, Berlin, Heidelberg.

Vygotsky, Lev S. (1967). *Play and Its Role in the Mental Development of the Child*. Soviet Psychology 5:6–18.

Walsh, D., Gentile, D., Gieske, J., Walsh, M. & Chasko, E. (2004). *Ninth Annual Mediawise Video Game Report Card*. Minneapolis, MN: National Institute on Media and the Family.

In: Mobile Learning
Editors: D. René and C. Aubin

ISBN: 978-1-53612-801-7

Chapter 2

MOBILE LEARNING SCENARIOS IN LANGUAGE TEACHING: PERCEPTIONS OF VOCATIONAL AND PROFESSIONAL EDUCATION STUDENTS

Ana Nobre[1,], PhD and Adelina Moura[2], PhD*
[1]Open University, Laboratório de Ensino a Distância e eLearning (LE@D), Lisboa, Portugal
[2]University of Coimbra, LabTE, Portugal

ABSTRACT

Mobile devices play a significant role in society, in general, and a very limited one at the different levels of education. Smartphones, tablets and other mobile devices allows learning to occur anywhere (and at) anytime. These powerful technological devices can enhance the teaching and learning processes by helping to promote collaborative and individual learning and broadening the boundaries of the classroom to different contexts of learning. Many students have mobile devices and their applications can provide access to learning outside the classroom, for greater flexibility and more dynamic learning. In this sense, the

* Corresponding Author E-mail: Ana.Nobre@uab.pt; adelina8@gmail.com.

articulation of technological and methodological efforts allowed us to create learning scenarios supported by the devices that students take to the classroom (BYOD), and use them to motivate and involve students in meaningful learning. These devices offer the advantage of integrating various technologies in the curricular contents, such as in foreign and mother language courses, representing a set of possibilities of ubiquity that can have great impact on the learning process. Thus, we developed strategies with vocational and educational students' methodologies, such as augmented reality, project-based learning, game-based learning, collaborative learning and gamification. In this text, we present the results of two mobile learning studies in teaching French as a foreign language (to 18-23 year-old-students) and Portuguese language, as a mother tongue (to 15-19 year-old-students), in vocational education, implemented as a mediation tool in education to promote the construction of learning and development of significant skills of collaborative work. From the data collection, through a questionnaire, with open and closed questions, we highlight the favorable perception of the students to the integration of mobile devices in learning, and the recognition of the benefits of the teaching strategies used throughout the year, in the increase of curricular learning.

Keywords: mobile learning, gamification, foreign languages, microlearning

1. Introduction

The development of mobile technologies in recent years has made possible changes in the way we work, buy, live and learn. This evolution has also been reflected in education, with some examples of the integration of mobile devices of learners or of schools / training institutions into educational practices. More and more students are using their mobile devices to support their studies, but most of the time they are banned from using them in the classroom. For this reason, schools should rethink their prohibitionist policies and adapt to the use of these technologies, accepting the new social, cultural and formative paradigm.

In the digital world everything changes rapidly and a constant update on the global state of the Internet is needed. According to recent data[6], more people are using the Internet than those who do not use it and the trend is for this number to rise. Regarding the use of mobile devices it already reaches more than 66% of the world population, and the rapid spread of smartphones has led to a significant growth of mobile Internet users globally. The same source, reveals that in Portugal 70% of the population uses the Internet, 59% use social networks, with mobile subscriptions exceeding (148%) of the country's total population and more than half of the population is active in some mobile device.

This evolution has had repercussions in the educational world, with an increase in pedagogical experiments using mobile devices (Han & Shin, 2016, Kukulska-Hulme et al., 2015, Moura, 2015a, Rikala, 2015, Sung et al., 2016). These experiments are helping the school to adapt to the needs of students, as they increasingly use mobile devices to support their studies, either by accessing learning content or as a working tool.

Mobile learning is a trend that is becoming more and more integrated into educational and training practices, a little bit everywhere, taking advantage of technological advances and social changes. Although there is no definitive definition of m-learning, several definitions have been proposed over the last decade showing the interest of the academic community in the subject. From the first definition suggested by Clark Quinn (2000)[7], very focused on technology, to the definition presented by Crompton (2013:4), in which she defines m-learning as "learning across multiple contexts, through social and content interactions, using personal Electronic devices," it is noted that the definitions have evolved over time, following the evolution of the technology itself and of its potential in the educational field. People are recognizing the benefits of mobile learning, in the educational field, enabling it to become, more like a standard approach and not an exception. For UNESCO (2013:5) "mobile technologies can

---

6 https://thenextweb.com/marketing-seo/2017/04/11/current-global-state-internet/#.tnw_CxzZ3TkA.

7 http://www.linezine.com/2.1/features/cqmmwiyp.htm.

expand and enrich educational opportunities for learners in diverse settings."

Since the first Mobile Learning Week in Paris, organized by UNESCO (2011)[8], in partnership with Nokia, which discussed how mobile technologies can be used to help achieve all educational goals. This year[9], the theme was "Education in emergencies and crises" and the participants discussed how existing technologies can strengthen inclusion in education, preserve continuity of learning in conflict and disaster contexts, catalyze innovation in the education sector and improve the impact of humanitarian interventions. Other events on mobile learning took place this year, showing the importance of this concept in education, such as The 4th International Mobile Learning Festival[10], where participants explored concepts, practices and issues that provide avenues for effective integration of mobile technologies and emerging learning technologies into education at all levels and in all environments.

The growing interest of academic research in integrating mobile technologies into education, in different educational contexts (Han & Shin, 2016, Kukulska-Hulme et al., 2015, Moura, 2015a, Sung et al., 2016), also reveals positive and interesting results, from the pedagogical point of view, and shows how much this area of knowledge still needs to be studied. As regards language learning different studies integrating mobile technologies have been developed in the past few years. Kukulska-Hulme et al., (2015) present a guide based on the research they have carried out, with experiments of teaching and learning the English language in different contexts, presenting the pedagogical framework for mobile assisted language teaching and learning[11].

M-learning experiments have been developed in formal and informal contexts and in collaborative learning environments (Crompton 2013). We present educational practices developed in classes of mother tongue

[8] http://www.unesco.org/fileadmin/MULTIMEDIA/HQ/ED/ICT/pdf/UNESCO%20MLW%20report%20final%2019jan.pdf.

[9] http://en.unesco.org/events/mobile-learning-week-2017.

[10] http://imlf.mobi/.

[11] Mobile Assisted Language Learning (MALL).

(Portuguese) and foreign language (French) with students of vocational and professional education.

2. Mobile Learning: The Importance of Technology in Schools

What is the real importance of technology in schools? What is its role and how can it effectively add value to the teaching and learning process? These are essential questions to better understand the transformative power of technology in education. We live in the age of knowledge and the skills and abilities needed today and in the future require collaboration, communication, creativity and critical sense. In the digital age in which we live, it is crucial to know what to do with content (which will always be important), how to produce, disseminate and share it, because the world is witnessing a massive production of information, such as never before in human history. How can digital mobile technology help us to stay connected at all times, to benefit from all the information? If we have mobile devices at hand, then it is necessary to strengthen and use these tools as best as possible to achieve the desired educational objectives, which are always the teachers priority and the school must always adapt to the new times.

A good pedagogical use of technology allows harmonizing pedagogy and technology in a good way, through practices and experiments typical of the times in which we live that require more flexible ways of thinking. But there are several factors that influence the inclusion of mobile learning experiments in formal education. Some of them are related to the integration of ICT in education, because mobile learning is part of educational technology (Rikala, 2015). Educational technology has evolved and spread rapidly but there is still no successful recipe. The truth is that there is not one recipe that fits all contexts. Therefore, it is best for each institution to look locally and discover the needs of its school community and adapt the technology integration plan to its own context.

The first step in creating a thriving technology ecosystem is students to have access to technology. This is already happening, when we adopt BYOD (Bring your own device) programs in our schools, because students come to school each time with more technologies, we just have to take advantage of them. But after having the technology, the teacher must create his or her own educational scenario, where students can explore, create and grow personally and intellectually, because we know that learning is much more than technology. It is about enabling students to develop the skills needed to meet the challenges they will encounter in future work.

Mobile learning is about creating the conditions for the students to create deep reflections and share them with others, for example using the class blog. Students can use the tablets or smartphones to search the places mentioned in the handbook or in a novel they are studying in class and share them in podcasts or in digital notes. Students from different classes can collaboratively develop a project that can help them solve a real problem. Students can create work groups and use online presentations to expound their ideas and search results. They can create short videos about the books they are reading, and QR codes with the links, and share them at school. They can share their work with classmates through Google Docs and get feedback from them. They can use Twitter to publish relevant information about a curricular topic and to comment on the publications of classmates, promoting social learning (Moura & Carvalho, 2013). Finally, mobile learning is about using Edmodo to work in groups and solve homework and generate discussion about a topic of the class. Technology must be at the service of learning, not vice versa. Therefore, it is necessary to start using it to learn how to enhance it and to benefit students' learning.

In a global and connected world, finding fixed and stable identities begins to be rare. The means and forms of online communication are constantly changing, requiring participants to develop new skills. Our young people also participate in this technological revolution, but often without the critical knowledge of research, the evaluation and information utilization skills. It is imperative that the school helps students develop digital literacy skills (Dudeney et al., 2013) to meet the challenges of today and the future. Digital competencies must be understood from multiple perspectives (Guikema & Williams, 2014) and can be developed in all disciplines and at all ages, because they are central to students' personal, social, educational and professional lives.

3. Strategies for Integrating Mobile Devices in Learning

To transform education with technology they are two ways, the pedagogical model and using tools to mediate the teaching and learning process. But how can we motivate students to learn? The students' lack of motivation towards curricular content is an obstacle to school success. The students of vocational education, coming from different contexts (social, family and school failure) become a very specific school population, requiring a constant readjustment of teaching strategies and curriculum plans from teachers. Therefore, there are many strategies that we use several times to make students become interested in the class subjects. For example, digital games, competition and rewards are strategies that have proved to be very positive at this level of education (Moura, 2015b).

Taking advantage of the mobile devices that students bring to the classroom, we have been able to implement a 1:1 model in our educational practices. Each student has their own mobile device (smartphone) and uses it individually. When the student does not have a device of his own, he can request one of the tablets from the school. Mobile devices are available and used whenever necessary. This situation helps to improve the management and organization of the class.

3.1. Participatory Culture: Learning Collectively

"Any reflection on the future of education and training systems in cyberculture must be based on a prior analysis of the contemporary mutation of the relationship with knowledge" (Lévy 1999: 157). The speed with which knowledge emerges and is renewed in all areas of knowledge is increasing and today, teachers and students, are constantly sharing, learning together, exchanging knowledge and producing knowledge.

This is evidently due to the advance of digital information and communication technologies, which have significantly changed the way we interact and understand the world. By changing the forms of entertainment, education, leisure, information dissemination and others, we are potentially changing all spheres of society and also, the way in which culture is understood today. A new culture is formed like this, the digital culture. According to Rüdiger (2013), cyberculture was historically explained "by the convergence of cybernetic thinking and communication computing."

Potentialized by digital resources, the school space has been updated. Now it is necessary to operationalize the changes that are taking place and to organize the new way of working with these innovations. Resources and pedagogical strategies are in full adaptation to this new context. Interaction and sharing become, increasingly, key words in the process of knowledge production. The studies of digital culture encompass all actions of production, application and expansion of data in cyberspace and also outside. In other words, cyberspace should not be seen as a parallel universe, not as an extension, but as part of our reality.

The student of today does not only want to consume information, he does not want to receive content passively, quite the contrary, he wants to produce knowledge. We are in participatory culture! For Jenkins (2008) there is content creation and sharing. Students are motivated because their contributions matter to others.

The characteristics of collaborative learning emphasize the autonomy of the student, and at the same time, value the collective construction of knowledge. That is, each one in his singularity can contribute to the growth of all by promoting reflections.

The highlight of the process is the common interest in the theme, the desire to make connections, establish relationships, contribute, be part of the group, and contribute to the collective, through the sharing of new knowledge. It is this that Gee and Hayes (2011) call the "space of passionate affinities" that people organize in the real world and / or via the internet (or a virtual world) to learn something about an enterprise, a shared passion or interest.

3.2. Learning: Student and Teacher (Co-Authors)

When dealing with the learning process in the school context, it is necessary to consider two extremely important actors: the student, as an active and participatory agent of the learning process, and the teacher as an agent in mediating between the student and the search for new knowledge.

Many teachers understand that the amount of classes is directly linked to learning, that is, "the more classes are given, the more the student will learn" (Demo, 2000). Knowing that the learning phenomenon of the students depends a lot on the teacher. A well-managed school with all the pedagogical resources is not viable if the school community, especially the teacher, does not leave aside the "traditional" classes and begins to consider learning to learn. The learning process must occur concomitantly in the student and the teacher, according to Demo (2007) "If we want to improve student learning, we must promote teacher learning." Regarding this statement, it is important that the teacher transmits to the students mechanisms of autonomy that they can develop in order to learn. In this context, the teacher needs to reconstruct his role in the teaching and learning process, to stop being the content transmitter and act as mediator, creating significant situations that allow students to assimilate knowledge. Demo (2007) points out that for the student to learn well, it is imperative that the teacher continues to learn well, that is, he is capable of provoking changes in the pedagogical procedures in the classroom, reflecting on the students' learning process, that is, he views himself as an eternal apprentice. Like Bourdieu (2000), we are convinced that... building a scientific object is first and foremost a break with common sense.

3.3. Audience Response Systems and Real-Time Polling

The web-based clicker tools are very popular with teachers and students. Research has been describing the benefits of using active learning approaches. Clickers or rapid audience response systems are a technology used to promote active learning. Martyn (2007) conducted a study comparing learning outcomes resulting from the use of clickers compared to active learning approaches to know concretely the cause of increased learning. Although without statistically significant results, both the use of clickers and the active methodologies help to improve students' learning.

Currently there are tools such as Kahoot, Socrative, Plickers, Quizlet, Quizizz, GoSoapBox that have replaced old clickers by taking advantage of the potential of smartphones or tablets with Internet access and that we integrate into our practices very often to learn, strengthen or consolidate vocabulary, grammar or literary content. One of the free online gaming-based learning platforms that can be used for any content area and easily integrates technology into your classroom is the Kahoot app[12]. Although this app only allows multiple choice answers and matching exercises, it is a good tool for reviewing and consolidating the subjects studied in class. Students love this game tool and it is a great way to integrate technology into the classroom and have students engage with curriculum content, while assessing knowledge in the digital age. We have seen using this app in our practices since 2014, possessing more than a hundred quizzes on the most diverse subjects of the program[13].

3.3.1. Assessment in the Digital Era

With the widespread use of smartphones among students and access to the Internet at any time, it was possible to provide students with learning and assessment of knowledge inside and outside the classroom. Using tools to create quizzes such as Kahoot, GoSoapBox, Socrative, Quizlet and Google Forms, students had the opportunity to solve quizzes and do tests in their spare time and in the classroom. These tools allow immediate

[12] https://kahoot.com/.

[13] https://create.kahoot.it/profile/linade.

feedback from students about their learning. Students use their smartphones to respond to real-time quizzes and polling. During this school year, we have integrated these tools into our practices, in order to involve and motivate students with their curricular content and to evaluate their knowledge. The data collected show that students have a positive opinion on using these tools for the success of their learning.

Since we have integrated digital technologies into more than ninety per cent of our classes, we have also changed student assessment tools, moving from paper-based tests to computer-based testing and rubric as assessment tools. Vis-à-vis rubrics, they allow the teacher to develop and systematize evaluation criteria and indicators, reducing the subjectivity of the assessment process. The use of rubrics makes possible to construct more transparent and coherent evaluation criteria in relation to the delimited learning objectives. The teacher can evaluate the students or groups of students with rubrics and the students can co-evaluate between them as well. The rubrics serve as formative assessment tools and allow students to be involved in the learning and evaluation process. Students can also make suggestions for the preparation of the rubrics through which the work and projects will be evaluated. For online-based testing we used Google Forms[14], and for rubric evaluation we used the public CoRubrics Google Drive templates[15]. We consider that these are useful and appropriate assessment tools to evaluate knowledge and skills in the mobile digital era in which we live.

3.4. *Diminuto* App: To Improve Reading and Writing Skills

Writing is one of the skills developed in the mother tongue class, but often the students are not motivated to write. So how can we improve writing skills in the mother tongue? One of the ways to improve writing skills in any language is to read actively in that language. Although there are different methods for improving writing, you need to have a broad

[14] https://goo.gl/forms/2t3MLnEWF7u57RM03.

[15] https://sites.google.com/site/corubricseng/.

knowledge of grammar and spelling. As the best writers are also good readers it is necessary to read regularly as it helps to gain confidence in reading and writing. Finding writing partners could be a good idea to engage students who do not normally like to write.

Based on these ideas, we presented to the students the App Diminuto[16]. Diminuto is a micro story app, up to 750 characters, in Portuguese language. It is possible to read the stories of other authors and start writing our own short stories. This app can be downloaded to iOS and Android devices. It's a productive hobby tool and a way to read a micro story in a minute. Some authors are also writers, others are just readers. We can create an account and write a short story that does not exceed 750 characters and send it for app author's approval. Then just wait for the app team evaluation result to see if it has been accepted or rejected. If it's published, it's available to read, to anyone accessing the app's website or on the mobile device. The creators of the app recognize their potential as a cultural hobby tool by enabling people to read short stories through mobile devices that fit perfectly into the routines of users and particularly our young generation.

As a motivation strategy for reading and writing, we present this app to students. After being installed on the students' devices, they had a week to read the published micro stories and share the stories preferred with classmates. This app has a simple and practical interface, which facilitates its use by the students and allows taking advantage of the free moments to read. In the second week students were challenged to create their own micro narratives outside the classroom individually. In one of the classes the proposal was the creation of texts in collaboration with another mate. This class was very dynamic and productive, because the students who had more difficulties were very committed to the writing of the collaborative stories. Some students expressed some difficulty in synthesizing the text to only 750-character and asked the teacher for help. This strategy proved to be very effective because it forced students to reduce the text to the essentials, which is not always an easy task. The imperative of brevity is an excellent exercise in writing and this app is situated for that.

---

[16] http://diminuto.me/home/landing.

Some students became enthusiastic and created several narratives. The students who wanted submitted their stories and some were approved, circulating now in the app. This possibility for students to see their published stories increased students' enthusiasm and motivation for reading and writing micro stories. This practice was also a great learning experiment for the teacher who also saw his micro narratives published.

3.5. Gamification: To Promote Students' Motivation

The concept of gamification was proposed in 2011, to designate "the use of game design elements in nongame contexts" (Deterding et al., p: 29). It has at its origin the idea of motivating, increasing the adhesion and levels of activity of users in tasks that are presented in an apparently playful format or with characteristics that resemble games or challenges.

The Foursquare location service, launched in 2009, was the application that gave the big boost to the idea of using game components where it would not be expected. The user, as he used the system, received points, medals, reached levels, etc., which made the service a huge success. The introduction of these game ingredients in a wide variety of fields, such as finance, health, education, among others, was quickly observed. At the same time, it began to be studied in the areas of man-machine interaction and game research (Deterding et al, 2011). Gamification relates to game, that is, presents elements that are characteristic of games. According to Deterding et al., (2011), what distinguishes the gamification of from games of entertainment or serious games, is that it is a system that includes some elements of games, but it is not totally a game.

According to Muntean (2011), gamification means introducing game mechanisms characterized by three attributes:

- Types of game mechanisms: Progression, feedback and behavioral;
- Benefits: *engagement*, loyalty, time, influence, fun;
- Personality types: exploiters, successful, social and murderous.

Thus, gamification is a learning strategy that takes advantage of the motivational elements of games in non-game contexts, such as educational and professional levels through the exploration of mechanical techniques (challenges, points, rewards,...) or dynamics (achievement, competition, incentive,...) in order to achieve better results. Why apply gamification in the classroom? Because it can improve students' learning outcomes and competencies at school. It allows rewarding the learning actions of the students and helps in the assimilation of knowledge.

There are more and more tools to gamifying the classes at different levels. Perhaps the best known are Kahoot, Socrative or Plickers. In addition to the use of these instant audience response systems we have also created the project "Magos da Sabedoria"[17] (Wisdom Magicians) to gamify the process of teaching and learning, available on the website shown on Figure 1. We intended to generate curiosity, expectation, overcoming obstacles, companionship and true emotions in our students. Besides that, we wanted to take advantage of the pedagogical potential of gamification to increase students' motivation and involvement in their own learning process.

Figure 1. Gamification project Website.

It is a Web environment in which a scenario of learning based on gamification is presented, motivating students to the study of two classic

[17] https://sites.google.com/view/magosabedoria

books of Portuguese literature, and working in pairs. This virtual learning environment is enriched with digital technologies. Students use their computer, tablet or smartphone, as being convenient for resolving different activities. This adventure is full of missions, challenges and rewards. Through videos, exercises, games, quizzes, challenges, evaluations and rewards students approach the study of two books that present some degree of complexity for young people. With this approach, it is a more fun and effective way to study curricular contents, considered difficult, while fulfilling the program of the Portuguese course and maintaining high levels of learning.

With gamification students are able to assimilate the most theoretical concepts at their pace and in a collaborative way. It creates a dynamic and interactive learning environment that helps to assimilate and creatively develop literary content. The students work teams discover the different missions and activities proposed according to their pace of learning. We believe that this strategy allows students to be responsible for their own learning and attends to the diversity of students' learning styles and increases the degree of motivation. The classes become more dynamic and fun. It was evident how the students' attitude evolved over the project, as they realized that the lesson evolved around their work, and not around a lecture by the teacher. This makes them responsible for the actions developed in each class, because they knew that the attitude, the effort and the interest in the class was rewarded in Leaderboard. In each class, the first reaction of the students was to see their position in the Leaderboard, reflected in the satisfaction of the great majority of the students for verifying that they were evolving and gaining points.

4. Methodology

The classroom experiments integrating the students' mobile devices were developed in the 2016/17 school year, at a public secondary school in Braga, in the north of Portugal, and at a Training Center in Lisbon. For the development of the different experiments that we presented above, we started from the idea that technology is increasingly present in society and in the educational world, so it is important to be present in classes too. This is a qualitative descriptive research.

To know the possession of the students' mobile devices and their use, we designed a questionnaire with multiple choice questions, in the Likert scale type, with three sections: personal information, possession of mobile devices, most used social networks and most used applications. The questionnaire was created with the Google Forms application. The first group was inserted into the class blog for students to respond, and the second was sent to the class by email. This questionnaire was introduced at the beginning of the school year, in September, to provide us with information about the possession of mobile devices by students and to help us consider the possibilities of integrating them into classes.

The first study comprises 66 students of the 10th grade, and 38 students of 11th grade, both from vocational courses, aged between 15 and 19 years old. The answers showed the prevalence of male students (86.5%)[18].

In the second study, 23 trainees took part in the professional course of the French-language class of a Professional Training Center, aged between 18 and 23 years old. These students, for the most part, had already dropped out of school some years before and did not have predefined curricular "paths," because "it is the students' abilities and needs that determine the curriculum to be considered" (Serra, 2005: 42), and the contents were presented in a more meaningful and motivating way. In order to stimulate a more personalized teaching approach, and more directed to the difficulties of each student/trainee, in the professional training courses, the classes have a smaller dimension than in the regular school, being constituted by 25 students at the most.

As we can see in the above data (Table 1) the smartphone is the most widespread mobile device among the students of the two groups, followed by the laptop. But, the habits of using multiple mobile devices, namely laptops and mobile phones are different. Another piece of data shows that most students in the 10th grade (78%) and 11th grade (80%) have their

[18] https://goo.gl/Nrdkhg.

own mobile data on their mobile devices. This data is important because it allows us to think about the integration of 1:1 student mobile devices in our practices, since the school has purchased tablets and can lend them to students who do not have a mobile device available.

Table 1. Students mobile devices ownership (N = 127)

Itens	**Vocational Courses**				**Professional Course**
	10 Grade (N = 66)		11 Grade (N = 38)		N = 23
	f	%	f	%	%
Mobile phone	18	30,5	16	42,1	15%
Smartphone	47	79,7	31	81,6	90%
Tablet	24	40,7	19	50	82%
MP3/MP4 players	7	11,9	8	21,1	21%
Laptop	35	59,3	32	84,2	87%
Other	2	3,4	3	7,9	----

The social network most used by all groups is Facebook, followed by Instagram and Snapchat. Most students access social networks on a daily basis. The mobile device most used by students in their private life is the smartphone, to access social networks, play games, send SMS, listen to music, communicate and search information on the Web. Most students in both groups (10th Grade 72.4%; 11th Grade 84.2%) consider that the students' mobile devices should be used in class as a tool to support learning.

Regarding students' perceptions of the potential of mobile devices to support learning activities, the vast majority of students in both groups (more than 80%) considered that the smartphone is an appropriate tool for reading PDF documents, access the Internet, take photos to illustrate work, make small videos, save notes, perform online tests, clarify questions with the teacher or classmates, organize field trips, and schedule events.

4.1. Student's Perceptions about Using Mobile Devices in Classroom

The data presented below were collected through an online questionnaire at the end of the school year, after the intervention plan was executed. We asked students in both groups about their perceptions of the mobile learning strategies used by the teacher during the year using their mobile devices (Table 2).

In the first group (Vocational Courses), for the vast majority of 10th grade (90%) and all 11th grade students using mobile devices helped them to understand lesson activities. For all students of both groups using mobile devices facilitated collaboration with other classmates and they enjoyed using the mobile devices to carry out the class activities. For 92% of 10th grade and 97% of 11th grade the use of mobile devices allowed them to be more participative in the development of tasks. Concerning class concentration using mobile devices, the vast majority of students in the 10th grade (89%) and 11th grade (95%) reported focusing more on learning when using mobile devices to accomplish the tasks of the lesson. This is relevant because mobile devices are often presented as distracting students (Rikala, 2015). All 10th grade students and 98% of 11th grade students reported that using mobile devices helped them with learning. All students in both groups considered mobile devices as learning tools. Finally, for 91% of the 10th grade students and 95% of the 11th grade using the mobile devices helped to understand the contents of the class.

In the second group (Professional Course), most of the trainees learned with mobile technologies because most trainees improved their French-speaking skills. All students liked to use the mobile devices for class activities. Most trainees have come to view mobile technologies not only as tools that help them in managing their school activities but also as valid pedagogical elements that are important to explore.

The totality of students from vocational courses enjoyed learning the lesson materials through Kahoot, Socrative, Plickers, GosoapBox and Quizlet, because they are more active and motivated in class, as pointed out in the open answers.

Table 2. Students perceptions about using their mobile devices in the classroom (N = 127)

Itens Scale		Vocational Courses				Professional Course
		10 Grade (N = 66)		11 Grade (N = 38)		N = 23
		f	%	f	%	%
Using mobile devices helped me understand lesson activities.	Strongly agree	19	61	26	32	2
	Agree	40	29	12	68	98
	Disagree	7	10	0	0	0
	Strongly disagree	0	0	0	0	0
The use of mobile devices in the classroom facilitated my collaboration with other colleagues.	Strongly agree	50	76	20	53	0
	Agree	12	18	18	47	100
	Disagree	4	6	0	0	0
	Strongly disagree	0	0	0	0	0
I enjoy using the mobile devices to carry out class activities.	Strongly agree	40	67	23	60	100
	Agree	20	33	15	40	0
	Disagree	0	0	0	0	0
	Strongly disagree	0	0	0	0	0
The use of mobile devices allowed me to be more participative in the development of tasks.	Strongly agree	51	77	23	60	2
	Agree	10	15	14	37	98
	Disagree	5	8	1	3	0
	Strongly disagree	0	0	0	0	0
I focus more on my learning when I use mobile devices to accomplish class tasks.	Strongly agree	30	45	19	50	--
	Agree	29	44	17	45	--
	Disagree	7	11	2	5	--
	Strongly disagree	0	0	0	0	--
The use of mobile devices helped me in learning.	Strongly agree	36	55	20	52	2
	Agree	30	45	17	45	98
	Disagree	0	0	1	3	0
	Strongly disagree	0	0	0	0	0
Mobile devices as learning tools help me in tasks.	Strongly agree	33	50	23	60	--
	Agree	33	50	15	40	--
	Disagree	0	0	0	0	--
	Strongly disagree	0	0	0	0	--
Using mobile devices helped me to understand the contents of the lesson.	Strongly agree	31	47	21	55	2
	Agree	29	44	15	40	98
	Disagree	6	9	2	5	0
	Strongly disagree	0	0	0	0	0

As we can see the perceptions of the students of both studies are very positive regarding all items, and about the potential of the mobile learning strategies developed throughout the year, supported by the students' mobile devices. This reveals that when properly used mobile devices and learning strategies can be very positive in students' motivation and interest in curricular learning.

CONCLUSION

As new tools and online services become available new opportunities emerge either for learning a foreign language or for improving one's mother tongue. The results show that using active strategies and mobile digital tools help students to better understand the lesson materials and to increase levels of participation and involvement.

Despite the students' perceptions about these experiences, there are constraints and challenges that the teacher has to solve during the learning process. In the past few years, we have been combining active methodologies and mobile technologies, and we have faced some obstacles that challenge the use of mobile devices, namely the Internet connection of the school that is sometimes slow and breaks the dynamics of the class, and this disappoints students. Also, performing some more complex tasks on a small screen is difficult for some students. Charging the students device batteries is another problem that arises, but this easily solved.

According to the results obtained, students of the two studies find more advantages than disadvantages in the integration of mobile devices and apps in the school activities and they complain that the integration of mobile technologies (devices and apps) in school should be greater. It is therefore important that teachers reflect on the possibility of implementing these tools in teaching practices in a natural way, for a better adaptation to the technological context in which we live, to enhance the functionalities that these tools bring to the teaching and learning process and to respond to students' needs.

We recognize the limitations of the data presented, collected and analyzed from a descriptive perspective, but we believe we can present strategies and ideas for a more efficient use of mobile technologies as tools of learning mediation in mother tongue and foreign language classes. It is necessary to continue to study the use of mobile devices in the practices of different school subjects, to obtain data that allow us to confirm trends or differences in this field. It might be interesting to compare these data with others obtained in other fields or levels of education and to analyze the use of mobile devices and applications in different educational contexts. By developing these experiments and integrating students' mobile devices into curricular practices, we are trying to make students more motivated in class, cause them to enjoy learning and to experience different technology-enriched learning practices.

REFERENCES

Bourdieu, P., (2001). *Langage et Pouvoir Symbolique*, Seuil. [Bourdieu, P., (2001). Language and Symbolic Power, Seuil].

Crompton, H. (2013). A historical overview of mobile learning: Toward learner-centered education. In Z. L. Berge & L. Y. Muilenburg (Eds.), *Handbook of mobile learning* (pp. 3–14). Florence, KY: Routledge.

Demo, P., (2000). *Conhecer e aprender: sabedoria dos limites e desafios.* Porto Alegre: Artes Médicas Sul. [Demo, P., (2000). Knowing and learning: limits wisdom and challenges. Porto Alegre: Artes Médicas Sul].

Demo, P., (2007). *Equívocos da educação*. Available from: http://pedrodemo.blog.uol.com.br/. [Demo, P., (2000). Misconceptions about education].

Deterding, S., Gixon, d., Khaled, R. & Nacke, L., (2011). *From game design elements to gamefulness: Definig "Gamification."* Proceedings from MindTrek '11. Tampere, Finland: ACM.

Dudeney, G., Hockly, N. & Pegrum, M. (2013). *Digital Literacies: Research and Resources in Language Teaching*. United Kingdom: Pearson Education Limited.

Guikema, J. & Williams, L. (Eds.) (2014). *Digital Literacies in Foreign and Second Language Education*. San Marcos, TX: Calico.

Han, I. & Shin, W. S. (2016). The use of mobile learning management system and academic achievement of online students. *Computers and Education*, 102: 79-89.

Kukulska-Hulme, A., Norris, L. & Donohue, J. (2015). *Mobile pedagogy for English language teaching: a guide for teachers*. Available from: http://oro.open.ac.uk/43605/.

Gee, J. P. & Hayes, E. (2011). *Language and Learning in the Digital Age*. London&New York: Routledge.

Lévy, P. (1999). *Cibercultura*. São Paulo: Editora 34. [Lévy, P. (1999). Cyberculture. São Paulo: Editora 34].

Martyn, M. (2007). Clickers in the classroom: An active learning approach. *Educause Quarterly*, 2, 71-74. Available from: https://goo.gl/8s6qGN.

Moura, A. (2015a). iPad Program in K-12 Education: The Pilot Year. In: ZHANG, Y. (ed.) *Handbook of Mobile Teaching and Learning*. Australia: Springer, pp. 601-616.

Moura, A. (2015b). Using arcade games to engage students in the learning of foreign and mother languages. In Giovanni Vincenti & James Braman (Eds.), *EAI Endorsed Transactions on e-Learning*, vol. 2, n. 5, pp. 1-14.

Moura, A. & Carvalho, A. (2013). Framework for Mobile Learning Integration into Educational Contexts. In Zane L. Berge, Lin Muilenburg (eds.), *Handbook of Mobile Learning*. London: Routledge, pp. 58-69.

Muntean, C. I. (2011). *Raising engagement in e-learning through gamification*. Paper presented at the 6th International Conference on Virtual Learning ICVL.

Rikala, J. (2015). Designing a Mobile Learning Framework for a Formal Educational Context. (Tesis doctoral). University of Jyväskylä, Findland. Available from: https://jyx.jyu.fi/dspace/bitstream/

handle/123456789/47324/978-951-39-6311-8_vaitos06112015.pdf; sequence=1/.

Rüdiger, F. (2013). *As teorias da cibercultura: perspectivas, questões e autores*. Porto Alegre: Sulina. [Rüdiger, F. (2013). The theories of cyberculture: perspectives, questions and authors. Porto Alegre: Sulina].

Serra, H. (2005). *Paradigmas da inclusão no contexto mundial*. Available from: http://repositorio.esepf.pt/handle/10000/27. [Serra, H. (2005). Inclusion paradigms in the global context.].

Sung, Y. T., Chang, K. E. & Liu, T. C. (2016). The effects of integrating mobile devices with teaching and learning on students' learning performance: A meta-analysis and research synthesis. *Computers & Education*, 94: 252-275.

UNESCO (2013). *Policy guidelines for mobile learning*. Available from: http://unesdoc.unesco.org/images/0021/002196/219641e.pdf.

BIOGRAPHICAL SKETCH

Ana Nobre

Affiliation: Universidade Aberta (Open University), Lisboa, Portugal

Education: PhD (Docteur d'État) in Didactics of Languages and Cultures, Université Panthéon-Sorbonne, Paris, France

Contact: Ana. Nobre@uab.pt

Research and Professional Experience:

Didactic E-learning; Open Education; Gamification in Higher Education.

After living and studying in Paris, she currently teaches at Universidade Aberta where she has taught since 1998, having previously been a Professor at the Sorbonne University, Paris. She completed a PhD in Didactologie des langues et des Cultures from the University Sorbonne Paris III. She is dedicated to the teaching of foreign languages in eLearning, to digital resources for learning in online environments and recently to the didactic eLearning and gamification in education. She was coordinator of the project "Teaching / learning languages online" and researcher of @ssess project of the Distance Education Laboratory and eLearning (FCT, 2010-2013) where she investigated the problem of digital alternative assessment of orality.

Some Publications from the Last 2 Years:

Articles in Portuguese language

Nobre, A., Mallmann, E., Recursos Educacionais Abertos: transposição didática para transformação e coautoria de conhecimento educacional em rede, *Indagatio Didactica*, vol. 8(2), julho 2016 Editora: CIDTFF - Indagatio Didactica - Universidade de Aveiro

Gomes, C., Pereira, A., Nobre, A., A gamificação no desenho personalizado no ensino superior a distância: um caso na aprendizagem de uma língua estrangeira, Universidade de Coimbra, *Faculdade de Psicologia e de Ciências da Educação*, LabTE, ISBN, 978-972-95595-9-4

Araujo, K., Nobre, A., *O Uso Inovador de Objetos de Aprendizagem sobre Ferramentas Informáticas de Apoio à Tradução Escrita no Âmbito do Ensino e Aprendizagem por Via Eletrônica para Docentes em Cursos Superiores de Línguas/Tradução*, 5o Seminário de Educação na Sociedade da Informação – Internet das Coisas: RESsignificando, Nova Friburgo, Centro Federal de Educação Tecnológica Celso Suckow da Fonseca, Brasil

Mazzardo, M., Nobre, A., Mallmann, E., Small Open Online Course com Professores do Ensino Médio: desafios para integrar REA nos materiais e atividades didáticas in V Congresso Brasileiro de

Informática na Educação (CBIE 2016) *Anais do XXII Workshop de Informática na Escola* (WIE 2016) Brasil.

Mazzardo, M., Nobre, A., Mallmann, E., Martin-Fernandes, I., Design-Based Research: desafios nos contextos escolares, *5o Congresso Ibero-Americano em Investigação Qualitativa* (CIAIQ2016), Porto, Portugal

Amador, F., Nobre, A., Barros, D., A Didática em context de eLearning: exemplos envolvendo coaprendizagem e coinvestigação in Digital Technologies & Future School, *IV Congresso Internacional TIC e Educação* 2016, Universidade de Lisboa

Pereia, H., Rosa, R., Nobre, A., REA*educa*- revista de Educação para o século XXI in Digital Technologies & Future School, *IV Congresso Internacional TIC e Educação* 2016, Universidade de Lisboa

Mazzardo, M., Nobre, A., Mallmann, E., Como aprender com recursos educacionais abertos? in *VII Congresso Mundial de Estilos de Aprendizagem, Instituto Politécnico de Bragança*, Bragança - Portugal

Book Chapters in Portuguese language

Amador, F., Nobre, A., Barros, D., Gonzaga, L., A Transposição Didática no âmbito do eLearning in Práticas e Cenários de Inovação em Educação Online, Universidade Aberta, *Educação a Distância e eLearning Nº1*, ISBN: 978-972-674-786-4.

Nobre, A., Mallmann, E., Quintas-Mendes, A., A Universidade Aberta a caminho da inovação pedagógica: recursos e práticas educacionais abertas, Práticas e Cenários de Inovação em Educação Online, Universidade Aberta, *Educação a Distância e eLearning Nº1*, ISBN: 978-972-674-786-4.

Cardoso, T., Gottsche, K., Machado, A. P., Nobre, A., Ensinar e Aprender Línguas Estrangeiras na Universidade Aberta: exemplos de práticas comunicativas online, Práticas e Cenários de Inovação em Educação Online, Universidade Aberta, *Educação a Distância e eLearning Nº1*, ISBN: 978-972-674-786-4.

Articles in international journals

Nobre, A., Pierre angulaire de l'enseignement des langues étrangères dans un environnement d'apprentissage virtuel: l'évaluation in Revue « *Contextes et Didactiques,*" n°9, Juin 2017, Antilles.

Nobre, A., Pereira, H., Rosa, R., a-REAEDUCA – Revista de Educação para o Século XXI: Pensar, Desenvolver e Criar um REA. *EaD em FOCO*, [S.l.], v. 7, n. 1, mai. 2017. ISSN 2177-8310.

Mazzardo, M., Nobre, A., Mallmann, E., Recursos Educacionais Abertos: acesso gratuito ao conhecimento? *EaD em FOCO*, [S.l.], v. 7, n. 1, mai. 2017. ISSN 2177-8310.

Burgues, A., Nobre, A., Chenoll, A., El proceso de enseñanza-aprendizaje en contextos ubicuos y universitarios. Tres estudios de casos in *Virtualidad, Educación y Ciencia*, Año 8 - Número 14 - 2017 ISSN: 1853-6530.

Nobre, A., Mallmann, E., *Um canal aberto no ensino superior? MOOC e REA no mundo digital in Apertura*, Universidad de Guadalajara, México.

Mazzardo, M., Nobre, A., Mallmann, E., *Small Open Online Course* e Recursos Educacionais Abertos na Formação Continuada de Professores do Ensino Médio no Brasil, Indagatio Didactica, Editora: CIDTFF - Indagatio Didactica - Universidade de Aveiro.

Book Chapters in English language

Amador, F., Nobre, A., Barros, D., Towards a Model of a Didactics of eLearning: An Application to Education for Sustainable Development in *Handbook of Research on Engaging Digital Natives in Higher Education Settings*, Copyright: © 2016 |Pages: 415 DOI: 10.4018/978-1-5225-0039-1.ch019.

Nobre, A., Setien, A., Chenoll, A., Digital eLearning: the different Educational Methods in *Handbook Optimizing Instructional Design Methods in Higher Education*. (Awaiting publication).

Nobre, A., Nobre, V., Pereira, H., Rosa, R., Collaborative learning and co-authors students in online higher education: a-REA*educa in* Student-Centered Strategies to Transform Adult Learning in *Online Environments* (Awaiting publication).

Nobre, A., Mallmann, E., Nobre, V., Mazzardo, M., MOOC and OER: identity management in k *Emerging Trends, Techniques, and Tools for Massive Open Online Course (MOOC) Management*. (Awaiting publication).

In: Mobile Learning
Editors: D. René and C. Aubin
ISBN: 978-1-53612-801-7

Chapter 3

THE EFFECTS OF INTERNET AND MOBILE EXPERIENCES ON MOBILE LEARNING ACCEPTANCE IN HIGHER EDUCATION

Mohamed Sarrab[1], Hafedh Al-Shihi[2], Zuhoor Al-Khanjari[3] and Asharul Islam Khan[3]

[1]Communication and Information Research Center,
Sultan Qaboos University, Sultanate of Oman
[2]Department of Information Systems,
Sultan Qaboos University, Sultanate of Oman
[3]Department of Computer Science,
Sultan Qaboos University, Sultanate of Oman

ABSTRACT

Mobile devices and applications with Internet capabilities have improved the convenience of information accessing for community of higher education learners. The learners and instructors acceptance of mobile learning is critical to the successful adoption of mobile learning therefore it is important to understand the factors that affect learners' intentions to accept mobile learning. From technology acceptance perspective, this research investigates the effect of learner's internet and mobile experiences on mobile learning acceptance in Higher education of Oman and other Arab Gulf Region. This study focused on how the level of internet and mobile experiences might influence learners' mobile learning acceptance to create a personalized learning experience in different higher education providers. The study has been conducted for 4 months from 17 different Omani higher education institutes to analyze and evaluate the relation between learners' internet and mobile experiences and mobile learning acceptance. The results of total 806 learners and instructors participated in this research showed Internet and mobile experiences have positive impact on learners and instructors' willingness and acceptance of mobile learning in different higher education providers. The effort is part of Omani-funded research project investigating the development, adoption and dissemination of mobile learning in Oman.

Keywords: mobile learning, M-learning, internet experience, acceptance; higher education

INTRODUCTION

The Internet has become very important in different aspects of our live as an important information resource related to Human-Computer Interaction. Learning activities have been changed due to the fast development in internet technologies as well as human computer interaction and communication have been changed due to the rapid development in mobile technologies. One of the most important characteristics of portable or mobile devices is their mobility and remote access to required information resources Alzahrani et al. (2014). Mobility can be defined as the possibility of changing locations across several times, and addressing different content areas using moveable devices such as smart phones, tables, PDAs and laptops Sarrab et al. (2016 and 2013). One more important characteristic of mobile technologies is their availability and easily access. The rapid development of mobile and communication technologies have the abilities to change and improve the nature of current learning processes Sarrab et al. (2014). The use of mobile technologies in learning system improves the learning process to be as location independence process and one-to-one relation of the interaction time. Actually, the mobility of learning process is relatively new approach that provided to support and improve students and teachers' effort and performance Sharma et al. (2016).

Mobile learning is the use of mobile and portable devices in the learning process considering user (learners, instructors and administrators), device and learning contents mobility and wireless connection Sarrab et al. (2015). Mobile learning is concerned with the ability to offer learning and educational content on mobile or portable devices. Learning and educational content refers to digital materials contents those might include any form of media or knowledge presentation made available on mobile device. However, mobile learning using handheld computing devices is still in its infancy age considering both pedagogies and technologies aspects Al-Harrasi et al. (2015). In 2000 Quinn defined mobile learning simply as learning approach that takes place with the support of mobile computing devices Quinn (2000). In 2002 Farooq et al., have defined mobile learning as a process that uses the advantage of mobility and wireless technologies in the learning Farooq (2002). In 2005 Traxler defined mobile learning as any educational provision where the dominant or sole technologies are palmtop or handheld devices Traxler (2005). In 2015 Sarrab has define mobile learning as new learning approach using handheld devices considering the mobility and wireless connection attributes Sarrab (2015).

Regarding the adoption of M-learning approach, it is essential to explore effect of Internet and mobile experiences on users' acceptance and perception towards the potential mobile learning applications Khan et al. (2015). Duggan et al., in 1999, stated that previous research studies on the Internet usage have tended to focus on the technical aspects of E-learning, while the users E-learning perception and acceptance were not studied Duggan et al. (1999). This research study have taken deeply and closer step to understand the potential of the mobile technologies to improve learning and educational uses performance and achievement.

Table 1. An overview of M-learning definitions

Source	Definition
Quinn (2000).	E-learning using mobile devices: Palms, Windows CE machines and even digital cell phone.
Riggs and Mark (2001)	An approach to E-learning or just another channel that delivers the same materials
Milrad (2003)	E-learning using mobile devices and wireless transmission
Malley et al. (2003)	Any type of learning happens when learners are not at a fixed, predetermined location when learners consider mobile technologies advantages for different learning opportunities.
Trifonova (2003)	Any sort of E-learning that happens using mobile devices or in mobile environment.
Georgiev et al. (2004)	A new approach of E-learning having the capability to learn everywhere and every time using of portable or mobile devices.
Keegan (2005)	The provision of training, learning and education on PDAs/palmtop/handhelds and smartphones.
Brown (2005)	E-learning that specially uses wireless devices for content delivery and learning support
Traxler (2005)	Any educational provision where the dominant or sole technologies are handheld or palmtop devices
Doneva (2006)	A new type or next stage of E-learning using mobile or portable devices and wireless network technologies for learning and teaching.
Diana (2007)	The digital support of collaborative, communicative, investigative, productive and adaptive learning activities in different remote locations that provide a variety of contexts for the teacher to function in
Sharples et al. (2007)	Processes and practices of coming to know over conversations using multiple contexts between learners and personal interactive technologies
Ting (2007)	A new education process that utilizes mobile devices as learning tools to gain the required information and resources anywhere and at any time.
Cavus and Ibrahim (2008)	E-learning, bounded by different characteristics and capabilities including different network technology properties
Shon (2008)	Learning using mobile computing and wireless network technologies to support learners' mobility and nomadicity nature.
Ally (2009)	The using of mobile computing devices to interconnect with fellow learners, instructors or institution as well as to access learning materials.
Tracey (2009)	Intersection of E-learning with mobile computing, learning empowered using mobile phones, PDAs and other personal and portable devices.
Jin (2009)	Rudiments of ubiquitous learning
Jiugen et al. (2010)	Fourth wave of learning using the Internet
Sarrab (2015)	New learning approach using handheld devices considering the mobility and wireless connection

THEORETICAL BACKGROUND OF MOBILE LEARNING

In terms of theory, Naismith et al. claimed that mobile learning is associated with many different theories, and examined various mobile activities including constructivist, collaborative, and behaviorist categories Naismith et al. (2004). Traxler stated that mobile learning community is over simplifying mobile learning concept by considering only the importance of E-learning, however, Traxler considers the transferability issue Traxler (2009). Pask in 1975 and Brown in 1996 have shown that social constructivist and conversation theories facilitate individualized communication and, collaborative learning Pask (1975), Brown and Campione (1996). Table 1 listed some researchers' definitions of mobile learning.

The previous definitions described mobile learning as knowledge building through life situations and within various contexts, which is enabled by learner, contents and technology mobility. Mobile learning is often considered as a subset of E-learning and next generation of D-learning. Mobile learning might be referred as S-learning (smart device based learning) or mobile technology based learning. Moreover, U-learning (Ubiquitous learning) is a form of mobile learning, where learning environments can be accessed in different contexts and situations.

METHOD

Procedure and Instruments

A survey has been developed and validated to measure the effect of mobile and internet experiences on mobile learning acceptance among learners in different Omani universities and higher colleges. A total sample of 806 learners and instructors from different higher education institutes participated in this research. The research study has been divided into two phases. The first phase involved study introduction about study objective, scope and methodology. This phase also involved discussion about participators internet experience and mobile learning. This phase involved a detailed formal discussion that discussed the survey objectives and processes with all study participators. This phase was divided into three parts, one detailed the research objectives showing the importance of the experimental study and the second part discussed how internet and mobile experiences might affect mobile learning acceptance. The third part discusses the methodology or study procedures whereas the study steps were explained to the participants. The second phase involved interview and survey study that consists of three parts: participators demographic background, level of internet experience and mobile experience. The participators general demographic information asked every participator about his/her gender, age, education level, area of specialization, study type and income. The second part involved level of participators Internet experience and the third part examined the participators mobile experience. Finally, the study participators were asked about willingness to use mobile learning or mobile devices for learning purposes. The survey had a total of 37 questions and it contained a number of multiple choose questions. After the data collect process through interview and survey the data cleaning process has taken place to improve the quality of data by reducing data errors and improving their presentation and documentation. The data cleaning process has been used to determine inaccurate and incomplete data and then modify and correct the detected errors. The data cleaning process included format, presentation, completeness and reasonableness using experts' assessment.

Respondents

The survey gathered data from 806 learners and instructors came from 17 different Omani higher education providers. 37 survey items were developed on the basis of consultation of IT experts and rigorous literature review. There were 5 items represent demographic background of participators and 32 items were used to measure the relation between internet and mobile experience and the mobile learning acceptance among

Omani higher education instructors and learners. Some of the 32 items were measured using 5 point Likert scale with 1 = strongly disagree and 5 = strongly agree.

Table 2. Participators characteristic (n = 806)

Characteristic	Sub	Participators	Frequency (Percentage)
Gender	Male	319.1	39.6%
	Female	486.8	60.4%
Age	18-23	481.9	59.8%
	24-29	93.4	11.6%
	30-35	58.8	7.3%
	36-41	87	10.8%
	42-47	38.6	4.8%
	48+	45.9	5.7%
Education	Ph.D	86.2	10.7%
	MS.c	144.2	17.9%
	BS.c	189.4	23.5%
	Diploma	191.8	23.8%
	Higher/Secondary	182.9	22.7%
	Others	12.1	1.5%
Profile	Full time	486.8	60.4%
	Part time	63.6	7.9%
	Instructor	203.9	25.3%
	Others	51.5	6.4%
Income	No income	354.6	44%
	Less than 251	146.6	18.2%
	251-500	16.9	2.1%
	501-750	22.54	2.8%
	751-1000	58.8	7.3%
	1001-1250	71.7	8.9%
	1251-1500	38.6	4.8%
	1501+	95.1	11.8%

All respondents were participated from different regions of Oman using simple random sampling technique. The basic participators group information is presented given in Table 1. In this study, about 40% of 806 participators are Males and around 60% Females. Most of the participators are young learners as the majority (59.8%) of the respondents belonged to age group 18-23, followed by 11.6% and 10.8% belonged to the age group of 24-29 and 36-41 years respectively. Furthermore, almost the same 22.7, 23.8 and 23.5 of the participants were studying in higher secondary, general diploma and, bachelor degrees. From the income perspective of participators, a round 44% do not have any income and 18.2% getting less than 250 Omani Rail (RO) where 1 RO is 2.59 USD. The rest of the income distributions are all below 1500 RO except 11.8% are greater than 1501 RO.

In the demographic background part participators were asked about their gender. Were about 60% of 806 participators are Females and around 40% are Males. The result of the age distribution part indicated that most of the responses were young people with about 60% ranging from 18 to 23 years old and the lowest number of responses were between 42 to 47 years old. Moreover, about 60% of participators were full time students and 25% were instructors. From education distribution perspective, the diploma, higher, and BSc each represent about 23% of 806 responses and the rest were distributed between Ph.D 10.7%, MSc 17.9% and others 1.5%. In terms of income, a round 44% of responses do not have any income and 18.2% getting less than 250 RO. The rest of the income distributions are all below 1500 RO except 11.8% are greater than 1501 RO.

Findings and Interpretations

Based on the survey study items regarding the effect of internet and mobile experience on the acceptance of M-learning, the participators respond were Interpreted. The study provided a systematic insight into participators acceptance of M-learning by empirically analysing and assessing the participators experiences about Internet and mobile devices. The participators were inquired about their internet experience, were 97.9% of 806 responses have internet experience. Moreover, about 36% of participators stated that they use internet more than 4 hours daily.

Table 3. Aggregate responses of internet experience

Questionnaire item	**Aggregate responses (Frequency)**				
Have you ever used Internet?	Yes (97.7%) 787.5	No (2.3%) 18.5			
Which device do you use most often for internet accessing?	Desktop/PC (17.5%) 141	Laptop (31.5%) 253.8	Smart phones (47.7%) 384.4	Tablets (2.9%) 23.3	Other (0.8%) 6.4
How long do you access internet?	Less than 1 hour daily (7%) 56.4	1-2hours daily (29%) 233.7	3-4 hours daily (26.5%) 213.5	More than 4 hours daily (35.6%) 286.9	Other (1.4%) 11.2
Which of the following tasks do you accomplish on the internet?	Never	Rarely	Often	Some	Always
Searching information	(0.5%) 4.0	(0.032%) 0.25	(13%) 104.7	(0.33%) 2.6	(50.3%) 405.4
Playing games	(37.3%) 300.6	(31.4%) 253	(19.5%) 157.1	(6.8%) 54.8	(5%) 40.3
Watching Videos online	(4.7%) 37.8	(12.4%) 99.9	(30.3%) 244.2	0% 0	52.6% 423.9
Taking quiz	(12.4%) 99.9	(39.3%) 316.7	(33.1%) 266.7	(10.9%) 87.8	(4.5%) 36.2
Downloading	(29.3%) 236.1	(26.8%) 216	(25.9) 208.7	(10%) 80.6	(8.1%) 65.2
Uploading	(11%) 88.6	(19.8%) 159.5	(28.4%) 228.9	(23%) 185.3	(18.1%) 145.8
Social networking	(12.6%) 101.5	(14.3%) 115.2	(22.8%) 183.7	(22.5%) 181.3	(27.8%) 224
Communication	(4.9%) 39.4	(11.3%) 91	(21.6%) 174	(27.6%) 222.4	(34.2%) 275.6
Chatting	(15.3%) 123.3	(20.1%) 162	(20.3%) 163.6	(16.8%) 135.4	(27.5%) 221.6

Table 4. Aggregate responses of Mobile device experience

Questionnaire item	Aggregate responses (Frequency)							
Do you own a mobile device?	Yes (95.8%) 772.1	No (4.2%) 33.8						
Which of the following do you think is a mobile device?	Desktop (4.7%) 37.8	Laptop (27.8%) 224	PDA (6%) 48.3	Tablet (11%) 88.6	Smart phone (26%) 209.5	E-book (0.4%) 3.2	All of them (23.9%) 192.6	Others (0.4%) 3.2
Which of the following software applications used most often on a mobile device?	Limited period free (18%) 145	Open source (66.2%) 533.5	Purchased (15.9%) 128.1					
Which of the following tasks accomplished using mobile devices?	Never	Rarely	Often	Some	Always			
	Accessing internet	(2.1%) 10.4	(5.8%) 46.7	(14.5%) 116.8	(27.3%) 220	(50.3%) 405.4		
	Searching information	(1.3%) 300.6	(6.9%) 55.6	(20.9%) 168.4	(32.6%) 262.7	(38.4%) 309.5		
	Messaging (SMS, MMS)	(0.9%) 7.2	(9.1%) 73.3	(18.8) 151.5	24.3% 195.8	46.9% 378		
	Recording of video	(7.2%) 58	(25.8%) 207.9	(32%) 257.9	21.1% 170	13.9% 112		
	Taking photographs	(3%) 24.1	(9.2%) 74.1	(26.3) 211.9	(27.2%) 219.2	(34.4%) 277.2		
	Installation of software	(5.6%) 45.1	(19.6%) 157.9	(38.3%) 308.6	(21.8%) 175.7	(14.7%) 118.4		
	Calculator	(6.1%) 49.1	(27.3%) 220	(34.4.8%) 277.2	(19.8%) 159.5	(12.4%) 99.9		

Questionnaire item	Aggregate responses (Frequency)							
	Taking quiz	(25.2%) 203.1	(34.2%) 275.6	(24.9%) 200.6	(10.9%) 87.8	(4.9%) 39.4		
	Playing games	(23.2%) 186.9	(26.8%) 216	(22.6%) 182.1	(15.6%) 125.7	(11.7%) 94.3		
	Watching Videos	(30.6%) 246.6	(11.7%) 94.3	(29.8%) 240.1	(28%) 225.6	(0%) 0		
	Video conferencing	(28.1%) 226.4	(26%) 209.5	(24.2%) 195	(12.6%) 101.5	(9.1%) 73.3		
	Downloading	(12.3%) 99.1	(21.9%) 176.5	(27.4%) 220.8	(19.1%) 153.9	(19.3%) 155.5		
	Uploading	(15.5%) 124.9	(25.9%) 208.7	(30.7%) 247.4	(16.5%) 132.9	(11.4%) 91.8		
	Chatting	(8.9%) 71.7	(8.9%) 71.7	(19.1%) 153.9	(20.9%) 168.4	(42.2%) 340.1		
	Communication	(4%) 32.2	(11.1%) 89.4	(24.9%) 200.6	(27.8%) 224	(32.3%) 260.3		
	Social networking	(12%) 96.7	(9.1%) 73.1	(18.8%) 151.5	(24.3%) 195.8	(46.9%) 378		

The participators were also investigated about their experience in using mobile devices. In this regards, 95.8% of 806 responses have reported that they have owned mobile devices and have mobile devices experience including configuration, download and insulation. Most of the response were using their mobile devices to accomplish some tasks on the internet such as accessing internet, searching information, messaging (sms, mms), recording of video, taking photographs, installation of software, calculator, taking quiz, playing games, watching videos, video conferencing, downloading, uploading, chatting, communication and social networking.

Table 5. Aggregate responses of willingness of mobile devices usage for learning purposes

Questionnaire item	**Aggregate responses (Frequency)**	
Willingness to use mobile devices for learning purposes?	Yes (90.2%) 727	No (9.8%) 78.9

DISCUSSION

Based on the findings of the experimental survey study, it was discovered that more than 90% of survey responses shown keen interest and willingness to use mobile devices for learning purposes. The research initial findings on the demographic background part indicated that most of the study participators are from young generation with different level of education and including both full and part time students and instructors. Whereas, most of the participators have not income. The majority of survey participators have used internet for different purposes and most of them uses smart phones for internet accessing more than four hours daily. The responses indicate that they always access internet to search information, watch video online, communication, chatting and social media. However, most of responses reported that they rarely taking online quiz. From the mobile device experience perspective, more than 95% of participators owned mobile devices, however large number of the them reported that mobile device is only smart phones and a few of them think that desktop is mobile device as well. Most of the survey participators accomplish some tasks using mobile devices e.g., accessing internet, search information, messaging, taking photographs, chatting, communication and social media. While some participates stated that not very often they install software applications and rarely take quiz or play games using their mobile devices, more than that some of participators have never used mobile devices for video conferencing. Finally and most important, that participators were investigated about their willingness to use mobile devices for different learning purposes, were more than 90% indicated their enthusiasm and willingness to use mobile devices as learning tool. Moreover, most of participators believe that the use of mobile devices for different learning purposes is wide spread in formal education and learning systems because of many learning, educational and user behaviour related reasons, such as:

- Information availability (anywhere and anytime)
- Ubiquity
- Portability.
- Flexibility.
- Interactivity.
- Accessibility.
- Collaborative.
- Visualization.
- Centred learning.
- Learners time usage.
- Improve confidence.
- Different learning styles.
- Reduce communication barriers.

CONCLUSION

No doubt that Internet and mobile computing devices currently consume learners' attention and time more than ever before. Assessing the effect of Internet and mobile experiences on learners' acceptance of mobile learning in higher education is crucial but neglecting the potentials that internet and mobile devices offer especially in the learning process is a serious issue. Mobile learning is an innovative learning approach that purposed to use technologies improvement in Internet connectivity and mobility to provide learners with learning solutions that reflects their modern world today. This research, focused on assessing the level of Internet and mobile experiences as influencers to learners' mobile learning willingness and acceptance to create a personalized learning experience in different Omani higher education institutions. Based on the study result, it was found that more than 90% of participators are keen interested to use mobile computing devices as learning tool. The study found that the majority of the participators are young from earlier higher education years and most of them with no financial income. The study discovered that most of learners and instructors uses internet and mobile devices for different duties more than four hours daily. Finally and most important, that participators indicate they always access internet using mobile devices for different tasks e.g., search information, watch video online, communication, chatting and social media. Moreover, the study explored that most of learners indicate that the use of mobile computing devices for different learning purposes is wide spread in formal higher education. Instead of only emphasizing on traditional educational tools and banning internet and mobile computing devices from higher education schools, it is necessary to re-think education with the positive impact of internet and mobile device on the overall learners' performance. The study is a higher education oriented; thus, the findings should not be applied on mobile learning from younger age groups.

ACKNOWLEDGMENTS

The authors would like to express their acknowledgements to Ms. Ibtisam Al Shibli and Nabeela Pathala Badursha for their assistant research work

REFERENCES

Al-Harrasi, H. Al-Khanjari, Z & Sarrab, M. (2015), Proposing a new Design Approach for M-learning Applications, IJSEIA, *International Journal of Software Engineering and Its Applications*, 9(11), 11-24.

Ally, M. (2009), Mobile Learning: Transforming the Delivery of Education and Training, *Issues in Distance Education Series,* Athabasca University Press., Athabasca, AB T9S 3A3, Canada.

Alzahrani, A. Alalwan, N & Sarrab, M. (2014). Sarrab. Mobile Cloud Computing: Advantage, Disadvantage and Open Challenge, *Proceedings of the 7th Euro American on Telematics and Information Systems*, (EATIS 2014) Article No. 20 ACM New York, NY, USA.

Brown, A. & Campione, J. (1996), Psychological theory and the design of innovative learning environments: On procedures, principles, and systems. In 'In L. Schauble & R. Glaser (Eds.) *Innovations in learning: New environments for education*', Hillsdale, NJ: Lawrence Erlbaum Associates, Hillsdale, NJ: Lawrence Erlbaum Associates, 289-325.

Brown, T. (2005), 'Beyond constructivism, exploring future learning paradigm', *Education Today* 2, 1-11.

Cavus, N. & Ibrahim, D. (2008), `m-learning: An experiment in using sms to support learning new English language words', *British Journal of Educational Technology* 40(1), 78-91.

Diana, L. *Mobile learning towards a research agenda Pedagogical forms of mobile learning: framing research questions* WLE Centre, 2007, 1, 151-173

Doneva, R., Kasakliev, N. & Totkov, G. (2006), Towards mobile university campuses, in '*International Conference on Computer Systems and Technologies - CompSys-Tech2006'*, Vol. VI of 3, CompSysTech2006, University of Veliko Tarnovo, Bulgaria, pp. 1-6.

Duggan, A. Hess, B. Morgan, D. Kim, S & Wilson, K. (1999). '*Measuring students' attitude toward educational use of the Internet*, ERIC Reproduction Service No. ED 429 117.

Farooq, U. Schafer, W. Rosson, M. & Caroll, J. (2002). *M-Education: Bridging the Gap of Mobile and Desktop Computing*, Centre for Human-Computer Interaction and Department of Computer Science. Virginia Polytechnic Institute and State University, 1-2.

Georgiev, T. Georgieva, E. and Smrikarov, A. M-Learning - a New Stage of E-Learning International Conference on Computer Systems and Technologies - *CompSysTech*' 2004, Office of naval research international field office, 2004, 1-5

Jin, Y. (2009), Research of one mobile learning system, in International *Conference on Wireless Networks and Information Systems*, 2009. WNIS '09', IEEE Explore, Shanghai, pp. 162-165.

Jiugen, Y., Ruonan, X. & Jianmin, W. (2010), Applying research of mobile learning mode in teaching, in '*International Forum on Information Technology and Applications (IFITA)*', Vol. 3, IEEE Explore, Kunming, pp. 417-420.

Keegan, D. The Incorporation of Mobile Learning into Mainstream Education and Training. *Proceedings of mLearn2005-4th World Conference of mLearning*, Mobilearn, 2005, 1.

Khan, A. Al-Shihi, H. Al-khanjari, Z & Sarrab, M. (2015), Mobile Learning (M-Learning) Adoption in the Middle East: *Lessons Learned from the Educationally Advanced Countries, Telematics and Informatics*, 32(4), 909–920.

Milrad, M. Mobile Learning, Essays on Philosophy, Psychology and Education Mobile Learning, Challenges, Perspectives and Reality. *Passagen Verlag*, 2003, 25-38.

Naismith, L. Lonsdale, P. Vavoula, G & Sharples, M. (2004), *Report 11: Literature review in mobile technologies and learning*, Technical report, University of Birmingham, Birmingham, UK.

OMalley, C., Vavoula1, G. N., Taylor, J., Sharples, M. & Lefrere, P. (2003), *Guide lines for learning/teaching/tutoring in a mobile environment.,* Technical report, MOBIlearn, UK.

Pask, G. (1975), Minds and media in education and entertainment: Some theoretical comments illustrated by the design and operation of a system for exteriorizing and manipulating individual theses, in 'In R. Trappl & G. Pask' (Eds.). *Progress in cybernetics and systems* research', Vol. 4, Hemisphere Publishing Corporation., London, 38-50.

Quinn, C. (2000*), mlearning: Mobile, wireless, in-your-pocket learning*, Technical report, Line Zine, Line Zine.

Riggs, R. & Mark, V. (2001), *Programming for Wireless Devices with the Java 2 Platform, Micro Edition*, Addison-Wesley Longman Publishing Co., Inc., Boston, MA, USA.

Sarrab, M. (2015). *Mobile Learning (M-learning) Concepts, Characteristics, Methods, Components. Platforms and Frameworks*, Nova Science Publishers, New york, USA, 2015. ISBN 978-1-63463-342-0.

Sarrab, M. Al-Shihi, H & Rehman, O. (2014). Exploring Major Challenges and Benefits of M-learning Adoption," *British Journal of Applied Science & Technology*, 3(4), 826-839.

Sarrab, M. Alzahrani, A, Alalwan. N & Alfarraj, O. (2014). From T-learning into M-learning in education at the University Level: Undergraduate Students Perspective, *International Journal of Mobile Learning and Organisation*, 8(3/4), 167-186.

Sarrab, M. Alzahrani, A. Alalwan, N & Alfarraj, O. (2015). An Empirical Study on Cloud Computing Requirements for Better Mobile Learning Services, *International Journal of Mobile Learning and Organisation*, 9(1), 1-20.

Sarrab, M. Elbasir, M & Alnaeli, S. (2016). Towards a Quality Model of Technical Aspects for Mobile Learning Services: An Empirical Investigation, *Computers in Human Behavior*, 55(A), 100-112.

Sharma, S. Sarrab, M & Al-Shihi, H. (2016), Development and validation of Mobile Learning Acceptance Measure (MLAM), *International Journal of Interactive Learning Environments*, Taylor & Francis Group.

Sharples, M., Taylor, J. & Vavoula, G. (2007), 'A theory of learning for the mobile age', *The Sage Handbook of elearning Research*, London 1, 221-247.

Shon, J.G. m-Learning Trends in Korea. ISO/IEC JTC1/SC36 Open Forum, ISO/IEC, 2008, 1

Ting, R. (2007), The advanced mobile learning practices: Learning features and implications, in '*Seventh IEEE International Conference on Advanced Learning Technologies*', 2007. ICALT 2007. *IEEE*, Taipei, Taiwan, pp. 718-720.

Tracey, W. (2009). *Education Technology and Learning Environments in Higher Education* Peter Lang International Academic Publishers; First printing edition, 214.

Traxler, J. (2005) Mobile Learning: It's here, but what is it? *Interactions*, 9 (1) University of Warwick.

Traxler, J. (2009), Learning in a mobile age, *International Journal of Mobile and Blended* 2(1), 1-12.

Trifonova, A. (2003), *Mobile learning - review of the literature*, Technical report, University of Trento, Povo Trento (Italy).

In: Mobile Learning
Editors: D. René and C. Aubin

ISBN: 978-1-53612-801-7

Chapter 4

ASSESSING SELECTION CRITERIA FOR CHOOSING M-LEARNING WITHIN A BLENDED LEARNING COURSE IN AFRICA

Ronald Hoyt Robertson III
Health Management, University of Applied Sciences Neu-Ulm, Neu-Ulm, Germany

ABSTRACT

Mobile Learning (M-Learning) in sub-Sahara Africa has the possibility to support learners along their daily life as the implementation of smartphones in everyday activities like banking and transportation is already happening in East African countries. The here presented study used the Lazy User Model to identify in which situations in the daily life of an East African hospital worker, the usage of m-Learning provided a real benefit in comparison to using e-Learning technologies. As the Lazy User Model assumes the preferred technology of a user to access information is chosen based on the lowest amount of effort expected, the study identified reasons beyond this in the African setting. To assess this, a blended learning course was examined where learning was performed using e-Learning and m-Learning, both online and offline, whereas the course was wrapped in two face-to-face workshops. A interview guide was developed and on-site interviews have been conducted with hospital workers coming from different departments, who graduated from this course, living in Kenya and in Tanzania. The results showed that unlike the assumption of the Lazy User Model, other factors affect the opting of m-Learning over e-Learning namely security and connectivity. Factors coming from the different instructional design of these formats could not be determined in this study group. The results of this pilot study present a first assessment of identifying the most suitable format of learning materials for East African countries which play an important factor for improving both the educational sector and the provision of further training for hospital workers.

Keywords: mobile learning, blended learning, lazy user theory

INTRODUCTION

A properly functioning health system requires well managed hospitals and appropriately educated hospital staff which is a prerequisite in order to tackle the Sustainable Development Goal (SDG) number 3 “Good health and well-being” for African countries. This goal is successively connected to SDG number 4 “Quality Education” and mobile learning (m-Learning) especially can serve as one tool to successfully implement this goal (United Nations, 2015).

M-Learning is seen as an opportunity for African countries to widen and improve the delivery of education in different sectors as mobile phones are already a ubiquitous tool in daily life activities of many Africans especially in Sub-Saharan Africa (Brown, 2003). In the first quarter of 2015 industry analysts noted that for the first time, more smartphones were shipped to Africa to enter distribution channels than feature phones with limited functionalities (Trucano, 2016). The widespread distribution of smartphones in Africa offers a new set of possibilities when designing m-Learning courses. Whereas a course solely provided in an m-Learning or an electronic learning (e-Learning) format only has limited possibilities of reaching and retaining learners, they can when being integrated in a blended learning course format, which also includes face-to-face training, provide a better learning experience for students (Kaur, 2013).

Blended Learning always represents a mixture of different learning styles and technologies, so the question occurs which "blend" fits best for hospital workers being in an African context. To identify this "blend," the here presented pilot study aims at finding these selection criteria to improve the design of such courses. The course design presented in this chapter allowed the students to choose their preferred learning technology, while other distance learning courses in sub-Saharan Africa often provide the participants with the necessary learning technologies, such as smartphones or laptops, leaving the learners to have little agency when choosing a suitable tool to take part in a blended learning course.

Considering the size of the available study group, this study assumes the sole limitation, when choosing the learning materials, are their physical surroundings, while leaving out factors, which can be detected using more extensive information sciences theories like the Unified Theory of Acceptance and Use of Technology (UTAUT) as proposed by Vankatesh et al. (2003).

Emerging Role of Mobile Learning in Sub-Sahara Africa

According to the Ambient Insight Report about the development of the worldwide mobile learning market from 2012 to 2017, Africa, represented in revenue forecasts from fourteen African countries including Kenya and Tanzania, has the biggest five year compound growth rate (CAGR) with 38.9%. By 2017 the revenues are expected to reach 530.1 Mil. USD, while half of the examined African countries alone are expected to exceed the 38.9% aggregate rate (Ambient Insight, 2013).

Figure 1 depicts the main catalysts responsible for the high growth rate of mobile learning in Africa, whereas the three catalysts "Boom in Mobile VAS", "Smart Inexpensive Devices" and "The Leapfrog Effect" were seen as most relevant for this study and will be described in detail.

The first catalyst "Boom in Mobile VAS" describes the growth of mobile learning value-added services (VAS) which is content sold by telecom network operators, device makers and content suppliers to consumers and organizations and delivered via audio, SMS or Interactive Voice Response. Considering the low technical requirements for phones to support these m-Learning VAS, this trend is especially growing in low income countries.

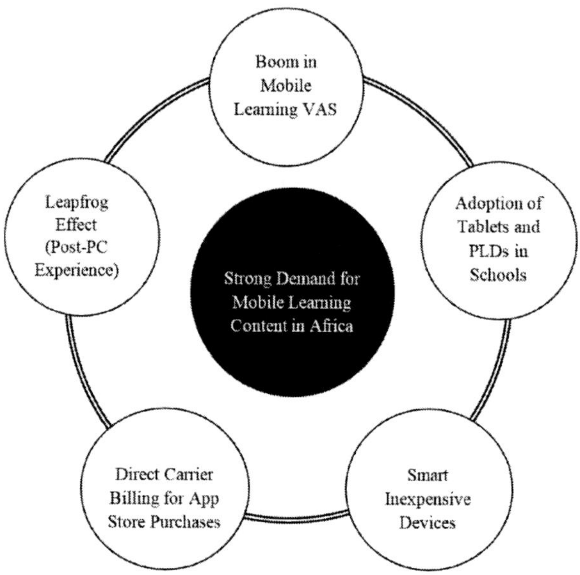

Figure 1. Primary Catalysts Driving the 2012-2017 Mobile Learning Market in Africa (Ambient Insight, 2013).

The most popular content delivered through this medium in Africa is Mobile Health (mHealth) and agro-culture educational content. Having 95 active mHealth programs, the African region, especially Kenya and Uganda were at the time of the report the most vibrant markets for educational content in the healthcare sector (Ambient Insight, 2013).

This catalyst is supported by initiatives by large corporations like Microsoft and Huawei who launched programs targeting the wide scale supply of low-cost smartphones for African youth (Stroh, 2013). The general decrease of the costs for smartphones, will lead to an expected growth of smartphone users in the Middle East and Africa from 85.7 Mil. in 2014 to 173.8 Mil. in 2019 (eMarketer, 2014).

The "Leapfrog Effect" describes the skipping of intermediary stages of technological development and thereby being able to directly access the latest mobile technology available such as 4G supporting mobile devices (Alaoui, 2014). Though being limited in its applicability in the African continent, mobile devices are the primary devices used for computing in many African countries and many children and adolescents in Africa will have their first experience with an IT device through a smartphone. Also the widespread acceptance of using mobile services to support everyday life activities such as banking, receiving payrolls, healthcare and everyday purchases through services like M-Pesa, largely supports this effect. Moreover this effect also determines the advancement of m-Learning and by the end of 2012, four African countries including Tanzania and Kenya were spending more on mobile learning than on Self-Paced Learning supported by e-Learning technologies. As a result, learners in Africa in the future will not substitute e-Learning content with m-Learning but are likely to leapfrog e-Learning altogether (Ambient Insight, 2013).

INVESTIGATED EXAMPLE

The course "Blended Learning - Hospital Management for Health Professionals (BL-HMHP)" was designed to suit the needs of African hospital workers from all departments with management responsibilities like doctors, pharmacists and chief nurses. The course content was created by German lecturers and the course was organized by the University of Applied Sciences Neu-Ulm.

Blended Learning Course Design

The design of the blended learning course will be illustrated based on a framework proposed by Stein and Graham (2014). It addresses the selection of learning technologies, the organization of learning activities, assessments, and support as well as the desired learning outcomes. To the existing framework, the topics "selection of learning technologies" and "Instructional Design" were added due to the fact that these design decisions and their impacts are the focus of the analysis later done in this study.

Course Overview

Hospital Management for Health Professionals (HMHP) is a ten-credit post-graduate short-course designed to suit the needs of African hospital workers in managerial positions and provides a foundation of basics in the area of hospital management. Although the course was designed for African countries, it was developed by a German lecturer team and was initially organized and coordinated from Germany. In later course iterations, the course coordination and organization was transferred to the African universities while the original content was kept.

Blended Rhythm

Being designed to fit in the African context by a German lecturer team, the rhythm of the blend is a framed model with onsite face-to-face seminars in the beginning and at the end, whereas the bulk of course activities are conducted online in between the two workshops in a six month online phase.

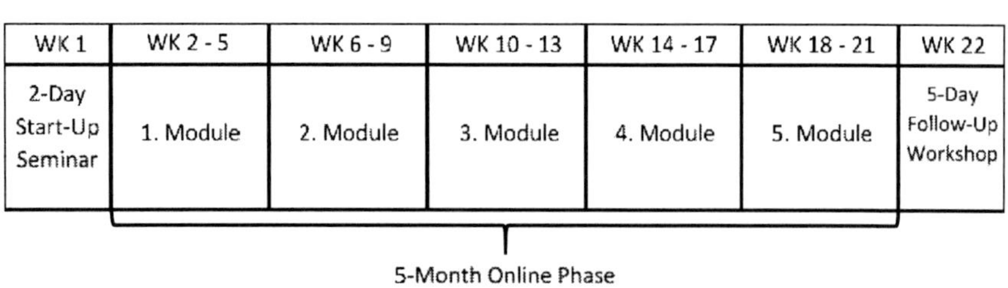

Figure 2. Framing Model of the BL-HMHP Course.

The course starts with a 2-day face-to-face start-up workshop in the respective home country of the participants where the course concept and the learning technologies necessary for the following five months were introduced. Afterwards the modules "Quality Management," "Health Economics and Material Management," "Human Resource Management," "Accounting & Controlling" and "Information Management" were taught in a 5-month online phase. During this phase each module was accompanied by a lecturer and tutor team to do regular follow ups on the learning progress of the participants, organizing and conducting chats, maintaining the forums and grading the individual and group assignments.

The course ended with a 5-day face-to-face follow-up workshop which program included a soft skills training, an excursion to a nearby hospital, a transfer project brainstorming session and an evaluation of the course.

Selection of Learning Technologies

Figure 3 summarizes the selection of the learning technologies now used in the HMHP course. The offline concept was a result of evaluations made which showed that participants of the HMHP course on average access 60% of the learning content offline.

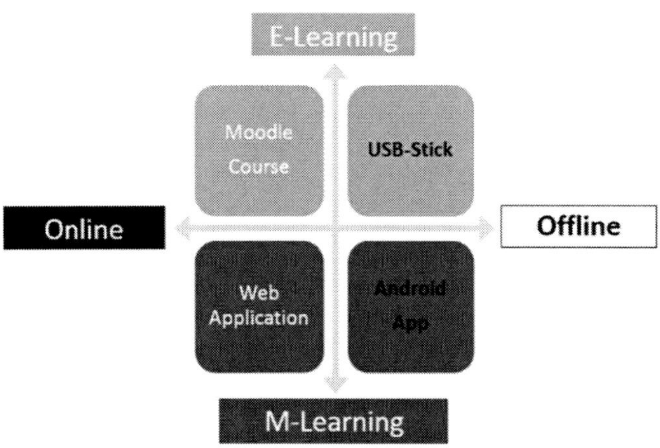

Figure 3. Learning Technologies in the HMHP Course.

In the first HMHP course conducted in 2013, the learning technologies used in the online phase consisted of an online course offered on Moodle and an offline offering with a DVD-ROM which contained solely the content of the five modules in a video and transcript version. As a result of the evaluation of this course, the demand for a mobile version was raised by students and was realized in a later course run in 2015. This mobile version consisted of an Android app which was installed on the devices of the participants and a mobile web-app which could be accessed over any browser. The Android version featured the content of the course in a non-animated and transcript based version to minimize the storage space required for installing and was suitable for any Android device with at least 50 MB storage space and an Android OS over Android 2.0 "Éclair". This ensured that any participant possessing an Android device could install the App without larger hard- or software restraints.

Instructional Design

The e-Learning content was created with the authoring system "Articulate Storyline" and included animated slides narrated by audio tracks. In the introduction of each unit a video from the lecturer was provided.

As seen in Figure 4 a slide in the e-Learning version was either animated or featured a video with animated content. On the left side the participant could quickly access a certain topic of the lesson and each lesson featured a small quiz or exercise to maintain the attention span during the 15-minute lesson. In total 60 lessons in this style could be accessed by the students either on Moodle or on the DVD-ROM.

Considering the design of content created in Articulate Storyline is suitable for regular sized desktop PCs or laptops, a new approach had to be used to design the learning content on the mobile version. As the mobile version should feature both a native app for Android smartphones and a web-app, the solution was found in a hybrid app. This app was developed using the jQuery Mobile framework and Phone Gap to transform the app later.

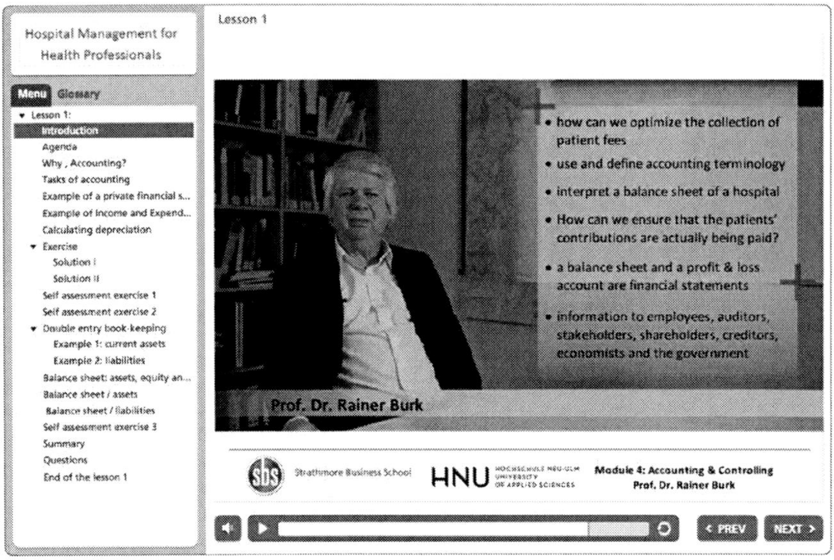

Figure 4. Screenshot from the e-Learning version of the HMHP Course.

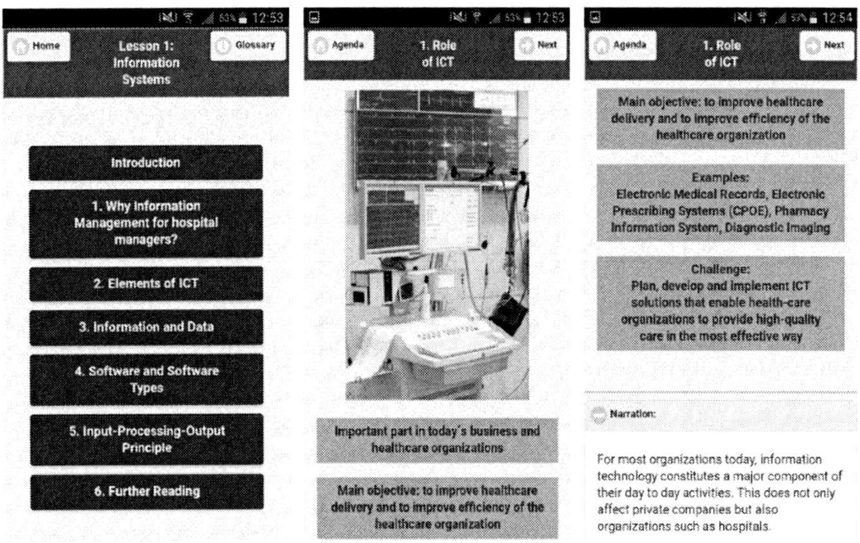

Figure 5. Screenshots from the m-Learning version of the HMHP course.

An example of the learning content design for the mobile version can be seen in the screenshots in Figure 5. The first screenshot from the left shows the navigation of the lesson with reoccurring items like "Exercises" and "Summary" highlighted. The second and third screenshot both show the slide "Tasks of accounting" while one shows the offline Android version with the narration in a collapsed transcript block below and the other shows the online web-app version with the narration as an audio track accessible over an audio player above.

Organization

The course concept of BL-HMHP developed over the years from its beginning in 2013 and progressed towards including more African lecturers and tutors from partner universities in Africa in the course organization. Therefore the concept described here covers the current concept where lecturers and tutors are from African universities and are trained by staff from the University of Applied Sciences Neu-Ulm (HNU) to conduct the course on their own.

Development of the Course IT-Infrastructure

The IT Infrastructure for conducting the BL-HMHP course can be seen in the image below.

For conduction the blended learning course HMHP with both an e-Learning and m-Learning offering, a university needs a Learning Management System (LMS) such as Moodle and a web server to host the mobile online application. The administrator of the LMS platform needs to create and setup a course room for all e-Learning activities, register all personnel and students involved and upload the required SCORM packages for hosting the learning content developed with the authoring tool Articulate Storyline. The mobile application will be uploaded as a regular website on a web server with password protection to hinder any unauthorized access. While Figure 6 describes the e-Learning and m-Learning online setup, the offline components consisting of a DVD-ROM or a USB-Stick and an Android app, were distributed in person to the participants on the first day of the start-up workshop.

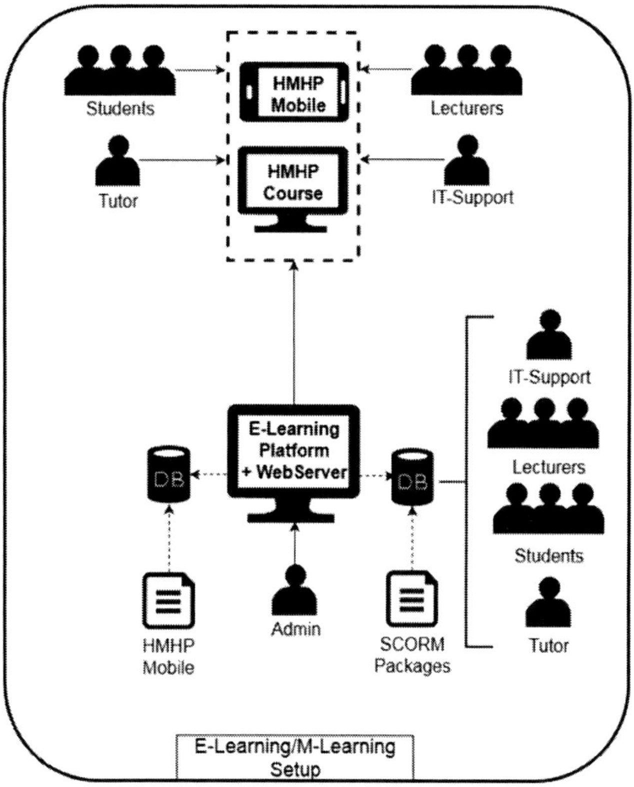

Figure 6. IT Infrastructure for conducting a HMHP course.

Conceptual Model and Hypotheses

For assessing the role of m-Learning within the presented example, the Lazy User Theory as proposed by Tétard and Collan (2009) was used. It assumes that a user who is seeking information, while having different options to access this information and depending on the state he or she is in, will always opt for the possibility, which requires the least effort.

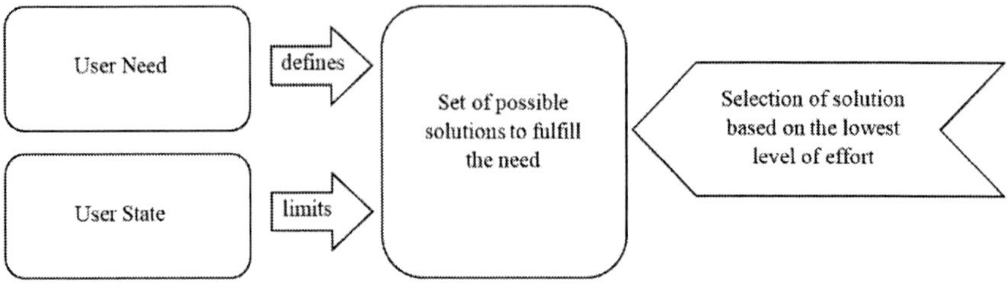

Figure 7. Lazy User Theory of Solution Selection.

The user state can be tangible or intangible and is in this case a piece of information which is specified in the type of information, the depth of information, the quality of information, the completeness of information and the urgency of information delivery. The user state limits the user due to the location he or she is currently in, the devices which are currently obtainable and the time available for attaining the information (Tétard & Collan, 2009).

From the Lazy User Model, the following hypotheses are derived:

- Hypothesis 1: A user need defines the set of possible alternatives
- Hypothesis 2: A user state limits the possible alternatives
- Hypothesis 3: The solution requiring the least effort from the user will most often be chosen

Applied to the mobile learning survey, the adapted lazy user model assumes the user need is learning the HMHP course by accessing the learning materials. This need is limited by three states, referring to locations the learner is currently in. Other factors like the available time and devices may derive from the three locations, "at workplace," "commuting to work" and "at home" as a result from the survey.

In the user state and user need scenario, as seen in Figure 8, the learner has three options of obtaining the desired information. He or she can use either a laptop or a PC to access the content of the HMHP course via Moodle or an USB Stick, or the information delivery is done via accessing the mobile learning system via either the Android app or the mobile web application. For each possible user state and possible solution, a four-point

Likert scale was used to measure the likeliness of the learner using each application.

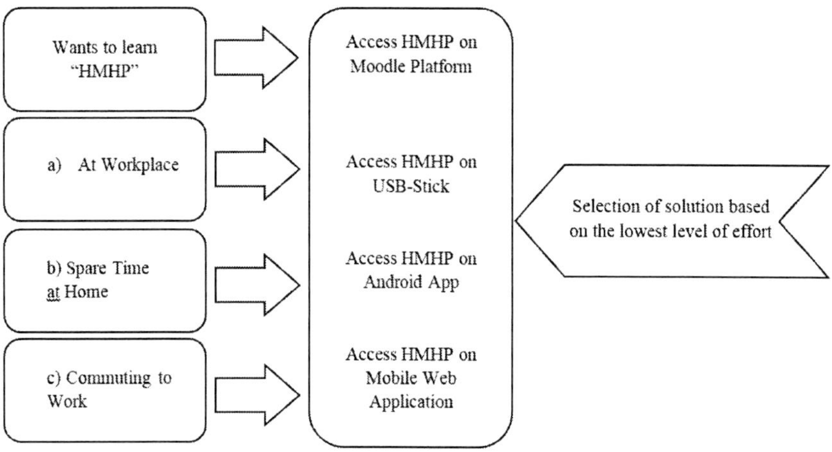

Figure 8. Lazy User Model applied in the Blended Learning Context of the HMHP Course.

METHODOLOGY

To better understand the context in which the research was conducted, this chapter will briefly describe the previous iterations of the BL-HMHP course which has been conducted in different African countries. While each iteration featured the same content, the organizational and technological components changed in each setting. The impact assessment done for this research focuses exclusively on the course iterations at Muhimbili University of Health and Allied Sciences (MUHAS) in Tanzania and Strathmore Business School (SBS) in Kenya as those were, at the time the research was conducted, the only settings in which the mobile offering was included. These examples show when m-Learning was used do to self-driven motivational factory in contrast to other study groups where a mobile phone or a laptop were the only available tools for accessing learning materials.

In order to analyse the impact of the mobile learning system in the blended learning courses at SBS and MUHAS, a mixed-methods study approach was chosen and therefore a survey was created which also served as a structured interview guide. This approach was chosen as it allowed surveying most of the former participants from the HMHP course while also allowing detailed insights into the responses from the participants. The survey was based on the Lazy User Theory adapted for mobile learning and was conducted both on-site as a structured interview and sent as an online survey.

From 36 eligible participants for the interviews/surveys, a total of 28 participated in the interviews and 2 participated in the online survey.

For the survey with the participants from Kenya, locations close to Nairobi were selected due to budgetary reasons. The interviews were conducted both in one-on-one interviews and in small groups of two to three participants at the hospitals where the participants were working and so Kenyatta Hospital, Mama Lucy Hospital and Westlands Health Centre were visited and at SBS as some of the participants were doing a Master course there. In total 15 former participants were interviewed, from which 7 used the mobile learning system and were eligible to participate in the Lazy User Model based questionnaire. As the survey taker was also the tutor from the HMHP course in 2015 and known by the participants, they were very welcoming and helpful during the interviews. The participants also knew, that the interviewer was the developer of the mobile learning system and therefore the Hawthorne Effect was taken into consideration before each interview. To minimize this effect certain counter measures as proposed by Gaskell were taken and so for example before each interview the "What's in it for me" was explained.

The participants of the MUHAS HMHP course were interviewed during a bus trip from MUHAS to the Tumbi Hospital. The interviewees, like in the SBS survey, knew the interviewer and his role in the App development and so the prevention measures to avoid the Hawthorne Effect were taken. Out of 12 participants from the course, 10 participated in the interview and 8 were eligible for the questionnaire. Due to the environment in which the interviews were taken, the amount of open answers and comments was limited.

RESULTS

For analyzing the results of the Lazy User Model based questions, the median and standard deviation were again taken to identify the most popular learning technologies in a given situation by the majority of the participants.

To better understand the following results, Table 1 shows the corresponding meaning of each Likert scale value.

Table 1. Meaning of Likert Scale Values

Value in Likert Scale	Meaning
1	Never Used the Technology in the Situation
2	Only Used Technology if Other Options Were not Available
3	Sometimes Used the Technology
4	Preferred Technology in the Situation

Studying HMHP at the Workplace

The acceptance of supervisors when learning with app was largely given, with some participants mentioning that they are their own supervisors and some stating that their supervisors encouraged them to take this training and where therefore interested in their progress and the content they were learning. One participant also said that even though using the smartphone was technically not allowed at work, it was still done as this regulation is hard to realize by the supervisors.

Table 2. Results of the Lazy User Model at the Workplace

Learning Technology	Mean	SD
Moodle Platform	3	0,93
USB-Stick/DVD-ROM	3	1,19
Mobile Web-App	1	1,1
Android App	4	1,13

As can be seen in Table 2, the Android app was according to the majority of the respondents the preferred option to study HMHP during work. When analyzing the interview responses from the work related questions, it becomes clear that the Android app was used as it presented the best option to learn in this situation rather than their preferred option which would have been in most cases the USB or DVD version.

Reasons why the Android app was the most suitable solution, were in many cases other programs running on the hospital PC and different users who had access to this PC, which did not allow for external resources like the HMHP course to be ran simultaneously. Other participants said that using the e-Learning was usually done in a comfortable time, whereas learning during work could most of the time only be done in short breaks where the mobile system was most fitting for the occasion.

Studying HMHP during the Commute to Work

During the commute to work, the clear choice was the Android app as most of the participants used public transport like buses or Matatus, which is a popular set of minibuses mostly found in Kenya, to go to work. In these cases and especially due to the limited space available for learning activities in crowded Matatus, the participants could not access their laptop for studying HMHP. When using the app to learn, the majority of respondents agreed to learn in the morning as in the evening the risk of their smartphones getting snatched from the windows of the buses was too high.

Table 3. Results of the Lazy User Model during the Commute to Work

Learning Technology	Mean	SD
Moodle Platform	1	0,96
USB-Stick/DVD-ROM	1	0,91
Mobile Web-App	1	1,1
Android App	4	1,12

Aside from the participants using public transport, the participants who drove to work themselves or by carpooling also had the chance to learn, because as most of them stated, due to long traffic jams a lot of idle time may occur during their commute.

Studying during Spare Time at Home

When studying during spare time at home, the participants could mostly refer to the learning technology, which they actually preferred without being constrained by other surrounding factors like being at work or commuting.

Table 4. Results of the Lazy User Model during Spare Time at Home

Learning Technology	Mean	SD
Moodle Platform	4	1,13
USB-Stick/DVD-ROM	4	0,99
Mobile Web-App	1	0,74
Android App	2	1,07

This can be seen in Table 4, as here unlike the other scenarios, the USB or DVD version were preferable. Although the results in Table 4 suggest that the Moodle platform was equally popular, it must be considered that many respondents saw Moodle as mandatory platform to visit as grading relevant activities like chats, forums and assignment submissions could only be done there. In rare occasions when the participants could not access their PCs or laptops, the Android app was still a fallback option for blackouts etc.

Other Scenarios in Which the HMHP Course Was Studied

The optional questions "Have you studied the HMHP course in a different scenario than those mentioned above?" was included to assess if all relevant learning scenarios possible for the learning groups were considered or to see if other options should have been considered in these cultures. In this section only isolated answers were given, like one female respondent who stated that she would suggest visits to the hairdresser are an often reoccurring scenario which is time consuming and was used by her to study the content of the HMHP course.

CONCLUSION

In this chapter the Lazy User Model was used to identify usage scenarios within East African blended learning courses where an m-learning application provided a benefit in comparison to the e-learning offering. Factors such as the limited degree of animation and audio or video material were less seen as a factor to limit the usage of the m-Learning offering in this study group but rather the screen size as well as security reasons during the commute with public transport. It was also noticeable, that any learning activity, which saw the usage of online functionalities of the LMS or the access to online resources as mandatory, were rather inconvenient due to the required journey to an internet café or the purchase of data bundles. M-Learning was used in scenarios where the learners would usually have to share their laptop or PC, like at the workplace or at home. While m-Learning had a significant impact on the learning experience of participants as it provided them with a better availability and flexibility to learn, this impact was limited to the offline version of the system. The current technological infrastructure, the willingness to pay for mobile data and the resulting attitude towards using an online m-Learning systems did not provide a suitable addition to the blended learning course environment presented in this example.

Considering the research done in this chapter would be widened to a larger cohort of hospital workers or other types of mobile learners in the Sub-Saharan African context, other information systemes theories like the Unified Theory of Acceptance and Use of Technology will be considered, to gather further insights on the self-driven motives of learners to use m-learning.

REFERENCES

Ambient Insight (2013). The 2012-2017 *Worldwide Mobile Learning Market*. Metaari, Washington: Ambient Insight.

Brown. T. H. (2003). The Role of M-learning in the Future. In: *21st ICDE World Conference.*

EMarketer (2014). *Number of smartphone users in Middle East and Africa from 2014 to 2019 (in millions)**. [Online] Available from: https://www.statista.com/statistics/494580/smartphone-users-in-middle-east-and-africa/ [Last Accessed: 18.02.2017].

Gaskell, A. (2012). How to Overcome the Hawthorne Effect [HTML]. [Online] Available from: http://www.peterborner.com/2012/03/07/how-to-overcome-the-Hawthorne-effect/ [Last Accessed: 18.02.2017].

Kaur, M. (2013). Blended Learning: Its Challenges and Future. *Procedia – Social and Behavioral Sciences, 93,* 612-617.

Stein, J., & Graham, C. R. (2014). *Essentials of Blended Learning: A Standards-Based Guide* (1st Edition). New York, NY: Routledge.

Stroh, M. (2013). *Huawei Unveils a New Windows Phone for Africa* [HTML]. [Online] Available from: https://blogs.windows.com/windowsexperience/2013/02/04/huawei-unveils-a-new-windows-phone-for-africa/#1rJHS65soHeTLPru.97 [Last Accessed: 18.02.2017].

Tétard, F. & Collan M. (2009). Lazy User Theory: A Dynamic Model to Understand User Selection of Products and Services. *Proceedings of the 42nd Hawaii International Conference on System Science* 2009, 1-9.

Trucano, M. (2016). *Reflections on The Last Five Years of 'Mobile Learning'* [HTML]. [Online] Available from: http://blogs.worldbank.org/edutech/reflections-last-five-years-mobile-learning [Last Accessed: 18.02.2017].

United Nations (2015). *Sustainable Development Goals [HTML]*. [Online] Available from: http://www.un.org/sustainabledevelopment/sustainable-development-goals/ [Last Accessed: 18.02.2017].

Venkatesh, V. et al. (2003). User Acceptance of Information Technology: Toward a Unified View. *MIS Quarterly Vol. 27 No. 3*, 425 – 478.

Biographical Sketch

Ronald Hoyt Robertson III

Affiliation: Department of Health Management, University of Applied Sciences Neu-Ulm

Education: Master of Science

Business Address: Wileystr. 1, 89231 Neu-Ulm

Research and Professional Experience: Research Assistant at the Department of Health Management at the University of Applied Sciences Neu-Ulm

Publications from the Last 3 Years:

Robertson, Ronald (2017) Entwicklung und Design einer Mobile-Learning-Applikation für die Schulung afrikanischer

Krankenhausmitarbeiter. In: *Digitale Transformation von Dienstleistungen im Gesundheitswesen II - Impulse für das Management.* Eds. Mario A. Pfannstiel, Patrick Da-Cruz, Harald Mehlich.

Robertson, Ronald and Gollmann, Manuel and Friedrich, Danny and Kania, Natascha and Kloß, Malte (2016) Meet Me, A Meeting Planner. In: *ICSE* 2016, Austin, Texas.

In: Mobile Learning
Editors: D. René and C. Aubin
ISBN: 978-1-53612-801-7

Chapter 5

APPLYING MOBILE TECHNOLOGY TO THE SUPPORT LEARNING AND DIAGNOSIS APPROACH IN A FLIPPED CLASSROOM

***Yen-Ting Lin*[1,*] *and Yi-Chun Lin*[2]**
[1]Department of Computer Science, National Pingtung University, Pingtung City, Taiwan (R.O.C.)
[2]Institutional Research and Development Center, National Kaohsiung University of Hospitality and Tourism, Kaohsiung City, Taiwan (R.O.C.)

ABSTRACT

Over the last decade, mobile technology has undergone an explosion and changed educational contexts. From out-of-class to in-class, the mobile technology provides high portability and mobility to facilitate teachers and students who develop mobile learning in various educational environments. In the mobile learning environment, the mobile technology can assist teachers and students in minimizing constraints of time and place and connecting real-world and digital-world resources effectively. In recent years, to facilitate the development of students' thinking, flipped classroom is an important pedagogy that reverses in-class and out-of-

[*] Corresponding Author E-mail: ricky014@gmail.com.

class learning activities in traditional classroom. In the flipped classroom, the unidirectional delivery of teacher's instruction in the traditional classroom is changed to deploy digital learning materials on the web and students can self-learn the instruction before participating the course. Moreover, the assignment of homework and practice out of class is changed to conduct in the classroom and the teachers and students can have sufficient interactions and discussions with regard to the assignment to facilitate thinking skills during the course session. As mentioned above, it is applicable to apply mobile technology to support the flipped classroom. In practice, the mobile technology has been widely applied to support out-of-class learning in the flipped classroom. However, it is rare to concern how to apply the mobile technology to support in-class learning in the flipped classroom. In this chapter, a cross-platform learning and diagnosis web application was developed to support a software engineering course in the flipped classroom. Moreover, a study was administered to investigate the effects on students' learning motivation and performance with regard to the use of the mobile technology in the flipped classroom-based software engineering course. The investigated results revealed that applying the mobile technology to support the software engineering course in the flipped classroom that is helpful to students in enhancing learning motivation and performance. In addition, most students showed favorable perception towards the usage of the proposed application.

Keywords: flipped classroom, software engineering, mobile technology, learning and diagnosis

INTRODUCTION

Over the last decade, mobile technology has a huge impact on education. From out-of-class to in-class, the mobile technology provides high portability and mobility to facilitate teachers and students who conduct mobile learning in various educational environments. Huang, Lin, and Cheng (2010) developed a mobile plant learning system to assist teachers and students in taking an outdoor learning for an elementary science curriculum in Taiwan. Ogata, Hou, Li, Uosaki, Mouri, and Liu (2014) proposed a ubiquitous learning system for capturing and reusing of learning log that aims to help students record, organize, recall and evaluate individual learning logs. Lin and Lin (2016) presented a mobile interactive learning and diagnosis system to support a clinical nursing course in class. Jou, Lin, and Tsai (2016) developed a theme-based mobile learning application to teach the practical techniques of material synthesis in class.

As mentioned above, the mobile technology can assist teachers and students in minimizing constraints of time and place and connecting real-world and digital-world resources in out-of-class and in-class learning environments. Therefore, several literatures proposed context-awareness mobile learning systems to detect the real-world learning status of students as well as the environmental contexts and further provide appropriate information to individual students in the right place and at the right time (Hwang, 2014; Hwang, Tsai, Chu, Kinshuk, & Chen, 2012; Hung, Yang, Fang, Hwang, & Chen, 2014; Wu, Hwang, & Chai, 2013). In the mobile learning environment, the learning system is usually able to offer supplementary learning materials or feedback to students based on their learning status. Moreover, the system can be used to record the students' real-world learning status as well as the interactions between the students and the system and further present appropriate learning guidance to the students based on the system analysis results (Lin, 2016). Therefore, the mobile technology is suitable to intervene various learning environments.

In recent years, to facilitate the development of students' thinking, a novel pedagogy, named flipped classroom, was proposed in 2007 (Chen, Wang, Kinshuk, & Chen, 2014). The aim of the flipped classroom is to reverse in-class and out-of-class learning activities in the traditional classroom approach. In the flipped classroom, the unidirectional delivery of teacher's instruction in the traditional classroom is changed to deploy digital learning materials on the World Wide Web and the students can use personal devices to self-learn the digital materials out of class before participating in the course. Moreover, the assignment of homework and practice out of class in the traditional classroom approach is changed to conduct in class and the teachers and students can have sufficient interactions with regard to the assignment to facilitate thinking skills during the course session.

To date, the flipped classroom has been applied in various educational degrees and courses (Slomanson, 2014; Teo, Tan, Yan, Teo, & Yeo, 2014).

Schultz, Duffield, Rasmussen, & Wageman (2014) applied the flipped classroom to support high school chemistry education. Kong (2015) implemented the flipped classroom to support an integrated humanities subject in a secondary school and further engage students in online pre-lesson learning preparation, in-class group discussion inside digital classroom and after-class extended learning using social learning platform. Lai and Hwang (2016) proposed a self-regulated flipped classroom approach to support a mathematics course in an elementary school. They also proposed an interactive e-book-based flipped learning approach to support a math course in an elementary school (Hwang & Lai, 2017). As mentioned above, to support the in-class learning activities in the flipped classrooms, the use of the mobile technology is still rare concerned in the literatures.

In this chapter, an investigation was conducted on using mobile devices to support in- and out-class learning activities in a flipped classroom-based software engineering course. Moreover, the effects of the proposed approach on students' learning performance were presented in this chapter.

Theoretical Foundations Underpinning This Chapter

The Flipped Classroom

In recent years, the flipped classroom has been applied in various subjects and several researches have been proven the positive effects of the flipped classroom on students' learning performance. Nevertheless, Kim, Kim, Khera, and Getman (2014) indicated that students' self-learning performance out of class is important to the flipped classroom since it affects how teachers and students interact with each other and learning materials in class. In other words, if the students do not have ideal self-learning performance out of class, then there is a strong risk that they may build relevant knowledge on faulty foundations in class (Lin & Huang, 2013).

In general, before conducting the flipped classroom, the teachers have to produce digital learning materials and deploy the learning materials on the cloud. The students can thus self-learn relevant learning materials on the web in the out-of-class learning activity of the flipped classroom. Following that, the teachers can develop meaningful activities to stimulate the students to engage in higher order thinking processes by utilizing the in-class time (Hao, 2016). However, in this learning mode, the students may lack learning motivation or may fail to schedule their time to complete the learning activity out of class. This phenomenon may also lead the teachers who are difficult to capture the students' learning status and prior knowledge level in class (Lai & Hwang, 2016; Lin, Lin, & Huang, 2011).

As mentioned above, therefore, to implement the flipped classroom, an appropriate learning and diagnosis approach is important to teachers and students.

Learning and Diagnosis

Evaluating students' prior knowledge level is a critical procedure before conducting relevant learning activities, since prior knowledge affects how teachers and students interact with each other and learning materials (Ozuru, Dempsey, & McNamara, 2009). Moreover, gaps in the students' prior knowledge often confound the teachers' performance to deliver instructions accurately. Furthermore, the students' prior knowledge also affects the teachers to plan instructional strategies for oncoming instructions. Therefore, it is important to evaluate how well students understand a concept or piece of knowledge prior to taking relevant learning activities.

To evaluate students' understanding on specific concept or knowledge, tests are usually adopted by teachers. In the past two decades, computer-assisted testing systems have been widely developed and proposed to evaluate students' concept or knowledge understanding level, because the kind of systems can provide a more convenient and flexible environment than is possible with paper-and-pencil tests (Huang, Lin, & Cheng, 2009).

Moreover, literatures have indicated that the assessment qualities of the computer-assisted test and paper-and-pencil test are equivalent (Cheng, Lin, & Huang, 2009; Chiou, Hwang, & Tseng, 2009). Nevertheless, the both testing approaches can usually use a score or grade to present the testing results that may be unable to provide further information to assist teachers and students in realizing what gaps or weaknesses may exist in the students' knowledge (Lin, Lin, & Huang, 2011; Wei, Lin, & Lin, 2016).

As mentioned above, therefore, several researches proposed novel assessment mechanisms to diagnose students' learning problems (Hwang, Panjaburee, Triampo, & Shih, 2012; Panjaburee, Triampo, Hwang, Chuedoung, & Triampo, 2013; Wang, 2014). Hwang, Panjaburee, Triampo, & Shih (2012) proposed a group decision approach to develop a testing and diagnostic system for diagnosing students' learning problems in mathematics. Panjaburee, Triampo, Hwang, Chuedoung, & Triampo (2013) developed a diagnostic and remedial learning system for an algebra course in a Thai lower secondary school based on the concept–effect relationship (CER) model. Wang (2014) proposed a two-tier diagnostic approach to implement a dynamic assessment system in an elementary school mathematics course.

Therefore, to assist teachers in conducting the flipped classroom and diagnosing students' prior knowledge, a learning and diagnosis system is developed based on the testing-based approach proposed by the authors (Lin, Lin, & Huang, 2011).

A Learning and Diagnosis Flipped Classroom Approach

To effectively support the flipped classroom, a cross-platform system is an essential element since teachers and students may use various learning devices to attend in- and out-class activities. Therefore, the development of a web-based system is an applicable cross-platform solution since the teachers and students can apply web browsers supported by any devices to use the system. Moreover, to consider different screen sizes, responsive web design (RWD) is also a necessary technique to develop the web-based system.

As mentioned above, a cross-platform learning and diagnosis web application was proposed in this study and deployed on a web server, as shown in Figure 1. By using the web application, the teachers are able to manage online learning resources and diagnostic assessments for oncoming courses. Moreover, the students can use personal learning devices with Internet to login to the web application through web browsers and further learn prior knowledge from the relevant learning materials before participating in the courses. Furthermore, the teachers can use the web application to diagnose the students' prior knowledge in class. Based on the students' learning performance out of class, the teachers can instantly adjust the teaching plan. During the course session, assume that the students have achieved a specific learning progress, the teachers can also ask the students to take a formative assessment to capture students' learning status through the use of the web application. According to the assessment logs, the web application can diagnose the students' learning problems and further provide the diagnostic results on individual dashboard to the students and teachers for improving their learning and teaching performance in class.

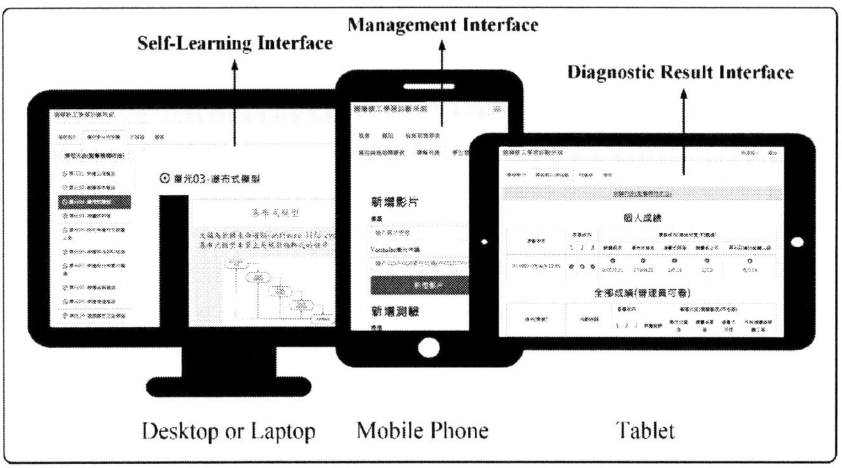

Figure 1. Screenshots of the cross-platform learning and diagnosis web application.

Software Engineering Education with Mobile Technology in Flipped Classroom

In this chapter, a study case in software engineering education was introduced to reveal how mobile technology can effectively support teachers and students to conduct the flipped classroom.

Experimental Design, Instruments and Procedure

To evaluate the performance of the proposed approach in software engineering education, a quasi-experiment was conducted on a flipped classroom-based software engineering course at a Taiwanese university. Participants in this investigation were 15 students (Male: 11 and Female: 4) and a teacher. All of the participants were supported by the flipped classroom strategy with the proposed web application to conduct the course. The course had a length of 4 weeks (10 hours).

In this experiment, a background knowledge test, learning achievement test, learning motivation questionnaire, and the perceived usefulness of the web application questionnaire were used to survey the students' learning performance and perceptions. The background knowledge test was used to evaluate the students' knowledge level with regard to the software engineering before taking the course. The learning achievement test was designed to assess the students' learning results after the conclusion of the course. The two tests included 10 multiple-choice test items respectively and the perfect score of the tests was 100 points. With regard to the learning motivation questionnaire, it was used from the intrinsic scale of motivated strategies for learning questionnaire (MSLQ) and consisted of nine items with a seven-point Likert scale (Pintrich & De Groot, 1990). In addition, the questionnaire for the perceived usefulness of the web application was used from the perceived usefulness scale of the technology acceptance model (TAM) (Davis, Bagozzi, & Warshaw, 1989).

To conduct this experiment, all of the students were firstly asked to take two pre-tests before undergoing the software engineering learning activities. The first pre-test was conducted to capture the initial learning motivation of the students by using the learning motivation questionnaire. The second pre-test was the background knowledge test to evaluate the students' background knowledge with regard to software engineering. Before taking the formal learning activities, the students received a 20-min instruction with regard to the flipped classroom strategy in the software engineering course. Following that, the teacher spent 30 minutes to instruct the operations of the web application to the students. In each week, the students were asked to conduct self-learning for learning theoretical concepts of software engineering by watching online video clips on the web application out of class. In class, each student was distributed a tablet to support the in-class learning activities. The teacher firstly applied the web application to capture the students' prior knowledge and determine the adjustment of the teaching plan. Furthermore, the teacher facilitated the students to engage in case study, discussion, and practice activities. In addition, during the course session, the teacher could also apply the web application to diagnose the students' learning problems and further discuss the diagnostic results with the students to promote learning. After going through the entire learning process, all of the students received a post-test with regard to the learning motivation by filling in the learning motivation questionnaire. Moreover, they also took the learning achievement test with regard to software engineering they learned in the course. Furthermore, they were further asked to complete a perceived usefulness questionnaire to survey their perceptions with regard to the perceived usefulness of the web application.

Experimental Results

The IBM SPSS was applied to analyze the collected data from the students in the experiment, including the results of the background knowledge test, learning achievement test, learning motivation questionnaire, and reception of the web application. To measure the students' background knowledge and learning achievements, two tests were conducted before and after the software engineering learning process. The descriptive statistic results of the two tests are shown in Table 1.

In order to investigate the effect of the web application on the learning achievements of the students in the flipped classroom-based software engineering course, a paired *t*-test was used to examine the difference between the background knowledge test score and learning achievement test score for the students. Table 2 shows that there was a significant difference between the students' test scores of the learning achievement and background knowledge ($t(14) = 10.693$, $p = 0.000 < 0.05$). The result reveals that the flipped classroom-based software engineering course with the web application benefited the students' the learning achievement.

With regard to the analysis of learning motivation, all participated students were asked to complete the learning motivation questionnaire before and after the learning process for evaluating their learning motivation. The pre-test and post-test Cronbach's alpha values of the questionnaire were 0.892 and 0.880, respectively. The descriptive statistic results of the two questionnaires are shown in Table 3.

Table 1. The descriptive statistic results of the background knowledge and learning achievement tests

Test	Number of students	Mean	Standard deviation
Background knowledge test	15	41.333	13.557
Learning achievement test	15	83.333	9.759

Table 2. The paired *t*-test results of the learning achievement for the students

Test	Number of students	Mean	Standard deviation	*t*(14)	*p*-value
Background knowledge test	15	41.333	13.557	10.693	0.000*
Learning achievement test	15	83.333	9.759		

*$p < 0.05$.

Table 3. The descriptive statistic results of the learning motivation questionnaire

Test	Number of students	Mean	Standard deviation	Cronbach's alpha value
Pre-test	15	5.311	0.651	0.892
Post-test	15	5.104	0.639	0.880

To further analyze the students' learning motivation, the paired *t*-test was used explore whether there was any significant difference between the means of learning motivation of the students before and after taking the entire learning process. Table 4 shows that the students' learning motivation was motivated after engage in the learning process ($t(14) = 2.225$, $p < 0.05$). From the statistical results of the learning motivation survey, it was found that the proposed approach promoted the learning motivation of the students.

To survey the perceptions of the students with regard to using the web application, the perceived usefulness questionnaire was used and revised based on TAM. The Cronbach's alpha value of the questionnaire items was 0.870, and the results of the investigation are shown in Table 5. The investigation results show that 84% of the students were in favor of using the web application.

Table 4. The paired *t*-test results of the learning motivation for the students

Test	Number of students	Mean	Standard deviation	*t*(14)	*p-value*
Pre-test	15	5.311	0.651	2.225	0.043*
Post-test	15	5.104	0.639		

*$p < 0.05$.

Table 5. The students' perceptions of using the proposed application

Question	Number of students	Mean	Standard deviation
Using the proposed application in the flipped classroom-based software engineering course would enable me to learn and diagnose relevant knowledge more effectively	15	6.066	0.593
Using the proposed application would improve my learning performance in the flipped classroom-based software engineering course	15	6.000	0.534
Using the proposed application in the flipped classroom-based software engineering course would increase my learning comprehension productivity	15	5.933	0.593
Using the proposed application would make it easier to learn software engineering in the flipped classroom	15	6.066	0.983
I would find the proposed application useful in the flipped classroom-based software engineering course	15	6.066	0.593

Discussion and Conclusion

In this chapter, a mobile technology supported flipped classroom learning and diagnosis approach was proposed to enhance software engineering education. An experiment was conducted in a university software engineering course to evaluate the effects of the proposed approach on students' learning performance. The experimental results showed that the proposed approach significantly improved the students' learning achievement and learning motivation. Moreover, most students agreed the usefulness of the proposed application in the flipped classroom-based software engineering course.

The results showed that the student participants achieved significant improvements in the learning performance after taking the flipped classroom-based software engineering course with the web application. The improvements were based upon several factors, such as the implementation of an appropriate pedagogical approach, the use of the mobile technology as well as well design and friendly learning application, the enough of the online learning resources, and the support of the learning and diagnosis services. To provide a comprehensive discussion, this chapter provided detail descriptions as follows.

To develop a flipped classroom learning environment, the effective application of technology is important to the flipped classroom (Bergmann & Sams, 2012). In this chapter, a well design cross-platform web application was used for facilitating content delivery and diagnosing learning status in the flipped classroom. Moreover, sufficient prior knowledge can also enhance students' learning performance in the flipped classroom (Lai, & Hwang, 2016). In addition, in a learner-centered learning environment, teachers play an important role to administer entire learning process. They have to encourage students to carry out discussions, express personal thoughts, and observed from the sides and gave feedback or assistance when required instead of giving a traditional one-sided practices or lectures. Finally, mobile devices also act an accelerator in the flipped classroom. The use of the mobile devices can break the layout of the traditional classroom that can effectively and efficiently promote teacher-student and student-student interactions (Lin, 2016).

In sum, the major contribution of this case study is to propose a cross-platform learning and diagnosis web application to support the mobile device-based flipped classroom in a software engineering course and further promote students' learning performance. Future directions of the issue would be focus on chatbot since chatbot is now a popular development and relevant technologies are ready to support the developments of mobile technologies. Therefore, to facilitate human-computer interaction in the flipped classroom, researchers can integrate chatbot techniques into learning applications to provide intelligent learning services for instructors and students.

REFERENCES

Bergmann, J., & Sams, A. (2012). *Flip your classroom: Reach every student in every class every day*. Eugene, OR: International Society for Technology in Education.

Chen, Y. L., Wang, Y. P., Kinshuk, & Chen, N. S. (2014). Is FLIP enough? Or should we use the FLIPPED model instead?. *Computers & Education*, 79, 16–27.

Cheng, S. C., Lin, Y. T., & Huang, Y. M. (2009). Dynamic question generation system for web-based testing using particle swarm optimization. *Expert Systems with Applications*, 36(1), 616–624.

Chiou, C. K., Hwang, G. J., & Tseng, J. C. R. (2009). An auto-scoring mechanism for evaluating problem solving ability in a web-based learning environment. *Computers & Education*, 53(2), 261–272.

Davis, F., Bagozzi, R., & Warshaw, R. (1989). User acceptance of computer technology: A comparison of two theoretical models. *Management Science*, 35(8), 982–1003.

Hao, Y. W. (2016). Exploring undergraduates' perspectives and flipped learning readiness in their flipped classrooms. *Computers in Human Behavior*, 59, 82-92.

Huang, Y. M., Lin, Y. T., & Cheng, S. C. (2009). An adaptive testing system for supporting versatile educational assessment. *Computers & Education*, 52(1), 53–67.

Huang, Y. M., Lin, Y. T., & Cheng, S. C. (2010). Effectiveness of a Mobile Plant Learning System in a science curriculum in Taiwanese elementary education. *Computers & Education*, 54(1), 47-58.

Hung, I. C., Yang, X. J., Fang, W. C., Hwang, G. J., Chen, N. S. (2014). A context-aware video prompt approach to improving in-field reflection levels of students. *Computers & Education*, 70(1), 80-91.

Hwang, G. J. (2014). Definition, framework and research issues of smart learning environments - a context-aware ubiquitous learning perspective. *Smart Learning Environments*, 1(4), doi:10.1186/s40561-014-0004-5.

Hwang, G. J., & Lai, C. L. (2017). Facilitating and Bridging Out-Of-Class and In-Class Learning: An Interactive E-Book-Based Flipped Learning Approach for Math Courses. *Educational Technology & Society*, 20 (1), 184–197.

Hwang, G. J., Panjaburee, P., Triampo, W., & Shih, B. Y. (2012). A Group Decision Approach to Developing Concept Effect Models for Diagnosing Student Learning Problems in Mathematics. *British Journal of Educational Technology*, 44(3), 453-486.

Hwang, G. J., Tsai, C. C., Chu, H. C., Kinshuk, Chen, C. Y. (2012). A context-aware ubiquitous learning approach to conducting scientific inquiry activities in a science park. *Australasian Journal of Educational Technology,* 28(5), 931-947.

Jou, M. Lin, Y. T., & Tsai, H. C. (2016). Mobile APP for motivation to learning: an engineering case. *Interactive Learning Environments*, 24(8), 2048-2057.

Kim, M. K., Kim, S. M., Khera, O., & Getman, J. (2014). The Experience of three flipped classrooms in an urban university: An Exploration of design principles. *The Internet and Higher Education*, 22, 37-50.

Kong, S. C. (2015). An Experience of a three-year study on the development of critical thinking skills in flipped secondary classrooms with pedagogical and technological support. *Computers & Education*, 89, 16-31.

Lai, C. L., & Hwang, G. J. (2016). A self-regulated flipped classroom approach to improving students' learning performance in a mathematics course. *Computers & Education*, 100, 126-140.

Lin, Y. T. (2016). When Mobile Technology Meets Traditional Classroom Learning Environment: How Does it Improve Students' Learning Performances?, In K. Wallace (Ed.), *Learning Environments: Emerging Theories, Applications and Future Directions* (Chapter 8). Nova Science Publishers, Inc.

Lin, Y. C., & Huang, Y. M. (2013). A Fuzzy-based Prior Knowledge Diagnostic Model with Multiple Attribute Evaluation. *Educational Technology & Society*, 16 (2), 119–136.

Lin, Y. T., & Lin, Y. C. (2016). Effects of mental process integrated nursing training using mobile device on students' cognitive load, learning attitudes, acceptance, and achievements. *Computers in Human Behavior*, 55(Part B), 1213-1221.

Lin, Y. C., Lin, Y. T., & Huang, Y. M. (2011). Development of a diagnostic system using a testing-based approach for strengthening student prior knowledge. *Computers & Education*, 57(2), 1557-1570.

Ogata, H., Houb, B., Li, M., Uosakic, N., Mouri, K., & Liu, S. (2014). Ubiquitous Learning Project Using Life-logging Technology in Japan. *Educational Technology & Society*, 17 (2), 85–100.

Ozuru, Y., Dempsey, K., & McNamara, D. S. (2009). Prior knowledge, reading skill, and text cohesion in the comprehension of science texts. *Learning and Instruction*, 19(3), 228–242.

Panjaburee, P., Triampo, W., Hwang, G. J., Chuedoung, M., & Triampo, D. (2013). Development of a diagnostic and remedial learning system based on an enhanced concept effect model. *Innovations in Education and Teaching International*, 50(1), 72-84.

Pintrich, P. R., & De Groot, E. V. (1990). Motivational and self-regulated learning components of classroom academic performance. *Journal of Educational Psychology*, 82(1), 33–40.

Schultz, D., Duffield, S., Rasmussen, S. C., & Wageman, J. (2014). Effects of the Flipped Classroom Model on Student Performance for Advanced Placement High School Chemistry Students. *Journal of Chemical Education, 91(9), 1334-1339.*

Slomanson, W. R. (2014). Blended learning: A Flipped classroom experiment. *Journal of Legal Education*, 64(1), 93-102.

Teo, T. W., Tan, K. C. D., Yan, Y. K., Teo, Y. C., & Yeo, L. W. (2014). How flip teaching supports undergraduate chemistry laboratory learning. *Chemistry Education Research and Practice*, 15(4), 550-567.

Wang, T. H. (2014). Developing an assessment-centered e-Learning system for improving student learning effectiveness. *Computers & Education,* 73, 189–203.

Wei, C. W., Lin, Y. C., & Lin, Y. T. (2016). An interactive diagnosis approach for supporting clinical nursing courses. *Interactive Learning Environments*, 24(8), 1795-1811.

Wu, P. H., Hwang, G. J., Chai, W. H. (2013). An expert system-based context-aware ubiquitous learning approach for conducting science learning activities. *Educational Technology and Society*, 16(4), 217-230.

Biographical Sketches

Yen-Ting Lin

Affiliation: Department of Computer Science, National Pingtung University

Education: PhD

Research and Professional Experience: Dr. Yen-Ting Lin is an assistant professor in the Department of Computer Science at National Pingtung University, Taiwan. He received his Ph.D. degree in the Department of Engineering Science from National Cheng Kung University, Taiwan, in 2010. He serves as a reviewer for three professional academic journals including Computers and Education, Computers in Human Behavior, and Interactive Learning Environments. His research experiences include educational technology, e-/ubiquitous/mobile learning, computer-assisted testing, expert system, and artificial intelligence. Dr. Lin has published more than 20 papers in such professional journals as Computers & Education, Educational Technology & Society, Interactive Learning Environment, and Computers in Human Behavior among others. He has 2 patents about educational technology.

Professional Appointments: Dr. Lin serves as a reviewer for three professional academic journals including Computers in Human Behavior, Computers and Education, and Interactive Learning Environments.

Honors: Dr. Lin received Benevolent Teacher Award from the Pingtung Educational Association in 2015.

Publications Last 3 Years:

1. Yen-Ting Lin*, & Yi-Chun Lin (2016). Effects of mental process integrated nursing training using mobile device on students' cognitive load, learning attitudes, acceptance, and achievements. *Computers in Human Behavior*, 55(Part B), 1213-1221. (SSCI, IF=3.435)
2. Min Jou, Yen-Ting Lin & Hsieh-Chih Tsai (2016). Mobile APP for motivation to learning: an engineering case. *Interactive Learning Environments*, 24(8), 2048-2057. (SSCI, IF = 1.674)
3. Chun-Wang Wei, Yi-Chun Lin, & Yen-Ting Lin* (2016). An interactive diagnosis approach for supporting clinical nursing courses. *Interactive Learning Environments*, 24(8), 1795-1811. (SSCI, IF = 1.674)
4. *Yen-Ting Lin* (2016). When Mobile Technology Meets Traditional Classroom Learning Environment: How Does it Improve Students' Learning Performances?,* In K. Wallace (Ed.), *Learning Environments: Emerging Theories, Applications and Future Directions* (Chapter 8). Nova Science Publishers, Inc.
5. Min Jou, Yen-Ting Lin, & Din-Wu Wu (2014). Effect of a blended learning environment on student critical thinking and knowledge transformation. *Interactive Learning Environments*, 24(6), 1131-1147. (SSCI, IF = 1.674)

Yi-Chun Lin

Affiliation: Institutional Research and Development Center, National Kaohsiung University of Hospitality and Tourism

Education: PhD

Research and Professional Experience: Dr. Yi-Chun Lin is a postdoctoral fellow in the office of institutional research at National Kaohsiung University of Hospitality and Tourism, Taiwan. She received her Ph.D. degree in the department of engineering science from National Cheng Kung University (NCKU), Taiwan in 2011. Her research interests are in institutional research, computer-assisted learning system, data mining and artificial intelligence. She serves as a reviewer for three professional academic journals including Soft Computing, Computers in Human Behavior, and Nurse Education Today.

Professional Appointments: Dr. Lin serves as a reviewer for three professional academic journals including Soft Computing, Computers in Human Behavior, and Nurse Education Today.

Publications Last 3 Years:

1. Yen-Ting Lin, & Yi-Chun Lin (2016). Effects of mental process integrated nursing training using mobile device on students' cognitive load, learning attitudes, acceptance, and achievements. *Computers in Human Behavior*, 55(Part B), 1213-1221. (SSCI, IF = 3.435)
2. Chun-Wang Wei, Yi-Chun Lin, & Yen-Ting Lin (2016). An interactive diagnosis approach for supporting clinical nursing courses. *Interactive Learning Environments*, 24(8), 1795-1811. (SSCI, IF = 1.674)

In: Mobile Learning
Editors: D. René and C. Aubin
ISBN: 978-1-53612-801-7

Chapter 6

MOBILE LEARNING AND SCIENCE EDUCATION

Tiziana Maria Sirangelo*
Department of Life Sciences,
University of Modena and Reggio Emilia, Modena, Italy

ABSTRACT

This study provides a critical analysis of the role that mobile technologies may play to support learning/teaching activities and in particular science education.

The opinions about their use in learning are different. Some are positive, and claim that mobile learning adapts well to the new pedagogic approaches of 'situated learning,' 'learning by doing' or 'authentic learning.' Others are more cautious and point out the almost total addiction of the youngsters to mobile technologies. Since the student uses these devices in everyday life to play and to communicate, it's unlikely he won't get distracted, preventing him from reaching concrete learning goals. Other experts insist that schools and more generally learning

* Corresponding Author E-mail: tizianamaria.sirangelo@unimore.it.

institutes should seize the opportunity to turn the means based on mobile technologies from distractions into learning tools. A means to do that could be integrating them in lessons and projects in the classes, providing similar learning context to the informal ones created by students in everyday life.

Developing an activity in which mobile devices are used means to elaborate learning processes where everyone, even in the collective context, may be partially protagonist. It's essential, then, to educate to a constructive use of such equipment, in which mobile technologies aren't only regarded as an entertainment or a way to communicate, but also as learning tools, a way to share, to collaborate and to support each other.

An extensive analysis of the inductive method in science education based on a student-centered approach has been carried out. It's an 'investigative' approach, in which the learner analyses the problem, observes, and formulates alternative hypothesis and comments in a critical manner. In other words, it's more like an active involvement of the student in the learning of sciences rather than a passive observation.

New technologies, including mobile ones, can't guarantee the achievement of satisfactory formative results if not coupled with well planned educational projects. However, in this context they may play a significant role and prove to be an effective support tool in several moments of sciences learning and teaching activities.

Keywords: mobile learning, digital technologies, teaching/learning methodologies, formative process, educational design

INTRODUCTION

M-Learning or Mobile Learning is the learning based on methods and materials that involve the use of mobile phones or handheld computers (smartphones, tablets, palmtops, and notebooks). It has enhanced upon e-learning by taking it a step further and allowing students to learn virtually anywhere a mobile signal is available. What's at the heart of this approach is that time and place constraints are totally absent (Sharples, 2013).

A significant number of ongoing studies about the adoption of school digital technologies and its development in the next years exists. They highlight how a computer represents both a learning and a research tool, as well as a good and simple way to put information together and get cultural and professional knowledge. These investigations focus on topics such as the integration between distance and conventional face to face teaching, and the educational aspects resulting from this hybrid approach. Innovative methodologies are explored as well. Projects, team work based on web and mobile technologies, virtual and remote laboratories are also interesting research topics and proved to be very useful to develop practice expertise before getting in real work contexts (Keengwe & Bhargava, 2014).

At the same time, the analysis underline how the diffusion of mobile devices has promoted social relationships, representing a good starting point, especially for youngsters. Therefore, young people can easily communicate with one another and can access to a wide variety of readily available game environments (Boyd, 2014).

The final goal of the recent studies and experimentations is ambitious and consists of verifying the potential of mobile tools not only to support educational and formative processes, but also to improve their quality.

First of all, in order to achieve such objectives it is necessary to define the teaching approaches to adopt and later to verify if the use of such technologies in school practice contributes to improve its development. Given these premises the central role of the educational and didactic design is evident.

It's also necessary to promote the acquisition of digital skills in youngsters, including technological, ethical and cognitive aspects.

Therefore, it is important to consider an alternative use of mobile devices, not only devoted to social communication and fun, but also to support the study of several subjects and generally the learning process.

However, teachers and researchers who want to adopt a m-learning approach have to take into account that experiences and professionalisms based on traditional digital laboratory are not easily adaptable to mobile technologies. In many cases some problems occur while trying to transfer teaching and learning methodologies and practices from computers to their personal and portable versions, such as a notebook (Kukulska-Hulme & Traxler, 2005).

M-Learning and Methodological Aspects

E-learning is based on desktop computer availability and therefore it can take place only in a given place, for instance in a classroom: this is, from a certain point of view, a limit. On the contrary, m-learning devices aren't dependent on a certain place to work. They can be used anywhere and at any moment. In addition, they can be kept in the palm of the hand and can be carried around quite easily.

However, mobile technologies can't be considered an alternative solution to the already available digital technologies, like whiteboards or computers. On the contrary, it is advisable that they interact with them, boosting their educational potentialities, instead of taking their place. As a consequence, they are regarded as a completion tool and an integration to the current technological equipment (Georgiev, Georgieva, & Smrikarov, 2004).

About their use in educational contexts opinions may be very different. Some of them do approve this new approach, like Prensky's one, according to which Mobile Learning fits the well situated learning, learning by doing and authentic learning pedagogical approaches, in both formal and informal contexts. Other opinions are more cautious and highlight the almost complete addiction of young people to mobile technologies.

Many researchers' concern is that the student, who uses the devices in everyday life for playing and communicating with his peers, will eventually get distracted and as a consequence will not achieve his learning objectives. Supporting that, when the student uses his mobile device individually he may prefer to look for leisure activities such as watching videos, surfing web sites or others different from the ones proposed by the teacher, who may lose instructional control.

Schools should seize the opportunity to turn mobile tools from distractions into learning means, by integrating them into the lessons in classroom projects and by recreating learning contexts similar to the informal ones spontaneously created by students in real life.

Developing a training activity based on mobile devices means to create an educational process in which each learner can be in a collective context and be individually protagonist at the same time.

Only a proper and careful design of educational activities will make such aims possible. For instance, challenging tasks or Internet based interactions aimed to stimulate the student's interest could be designed (Kukulska-Hulme, Sharples, Milrad, Arnedillo-Sánchez, & Vavoula, 2009).

The teacher's participation may contribute to improve learners' motivation besides having a control function. Tracking the learner's actions may increase his attention and involvement level. The so growing ability to produce and then show to other students his work may represent a rewarding and effective way to encourage a confrontation among different groups.

It is evident that the key concept to turn mobile technologies into a very successful educational tool is "integration". An effective project should not be developed in a context different from a basic school curriculum, but should be integrated into it and into students' experiences by enhancing their relationships with parents, teachers, other students and friends.

Provided this, it is believed that it is necessary to educate to a responsible use of these technologies. In fact, on the one hand, being constantly connected makes it possible to get support at any moment and being literally flooded with information. On the other, with headphones always on, teens live in a state of isolation from real world.

Finally, it is essential to promote a constructive use of such devices, in which they aren't considered just from a playful and communicative point of view but as tools devoted to learn, collaborate and support each other, if needed.

M-Learning and E-Learning

E-learning has completely transformed traditional learning activities and made on-demand knowledge available at any time by using web or intranet network. Nowadays the same contents adapted in a suitable way may be offered by using the mobile devices we all are familiar with, and this makes the learning experience even more effective and flexible (Georgiev, Georgieva, & Smrikarov, 2004).

In fact, due the relatively small size of such devices, mobile learning may be considered a tool devoted to provide information whenever and wherever needed. It has been proven to achieve the "learn by doing" paradigm better than the traditional learning that needs to take place in a classroom or in front of a computer.

During e-learning activities it is necessary to create a specific context to explain a given topic. During m-learning activities, learning takes place exactly when there is the right context. Therefore, one of the terms that characterize mobile learning is "contextual."

In addition, it is incredibly easy to login in downtime when using technologies such as smartphones and tablets, and gain access to, for instance, a short-lived training module.

This does not mean that a kind of learning is better than another one but both may answer the purpose in the right situation.

From this point of view, mobile devices are only another kind of tools which learners may use to access to their training materials.

Use of Mobile Technologies and Science Education

The use of innovative technologies in the education field is not necessarily successful, especially if they are not integrated with well processed training activities. When that happens, they play a significant role in projects related to laboratory practice (Hofstein & Lunetta, 2004), collaborative learning, individualized educational paths and research.

New technologies allow to redesign training plans and to implement a *student-centered approach*, aimed to develop a critical attitude and to promote collaboration among the learners. In fact, their use has changed the way students acquire information coming from alternative sources to traditional books and how they comment on the results of their research, by sharing opinions with other members of the scientific community.

In science education, a proper use of digital technologies is essential and many practical projects could be designed.

For instance, the *virtual laboratory* has proven to be a very useful scientific tool, making it possible to simulate real laboratory activities. In fact, it allows to perform online experiments of chemistry, physics and biology and it is a pretty easy environment to use for teachers and learners. In particular, a synchronous virtual laboratory is suitable for remote and interactive short-lived experiments. A remotely and real time controlled telescope is an example of synchronous laboratory. On the contrary, an asynchronous virtual laboratory is a more suitable choice for experiments which take a longer time to be carried out. A camera that takes a picture every hour is an example of equipment required for an asynchronous laboratory. Mobile technologies are more common than others and they facilitate laboratory practice, allowing remotely controlled experiments.

The so called *digital and mobile laboratories* are also commonly used, and transform a traditional classroom into a multimedia and interactive space in which several learning models may coexist: individual, teamwork, and so on. Notebooks are provided with special software which support science teaching and integrate the highest technological innovations with collaborative and laboratory methodologies. Learners are given BYOD devices (Bring Your Own Device). Then, they start observing, experimenting, measuring and eventually put data together by using the available technological tools. In addition, they can access online service supports for the studying at home. So the communications between school and family take place in a more direct way.

In this structured environment, an interactive training action, which is very different from traditional face to face teaching, takes place.

Alternative learning spaces are thus created, allowing to create a collaborative laboratory.

In order to achieve this aim, it is essential that the adopted methodological approaches are based on flexible guidelines, making it possible to involve students having different training needs and to overcome possible barriers that may prevent them from getting the best out of the educational dialogue.

Besides these "mixed" solutions, which make use of personal computers as well as of mobile devices, *laboratories based only on mobile technologies* exist, and allow the learners to engage themselves in experimental activities on the field (Duh, Tan, & Chen, 2006).

For instance, students may be provided with graphing calculators directly interfaced with sensors. This makes it possible to develop activities in small groups, with a greater involvement. At the same time, the increasing use of USB pendrives improves the flexibility of this kind of experimental lessons.

Portable tools, such as netbooks with an integrated USB microscope interfaced with different sensors, allow to design training activities devoted to research, observation of natural phenomena *on site*, and provide the advantage to have an immediate feedback.

Transferring the technology out of the classroom means to extend learning areas out of it as well, by offering the opportunity to develop a scientific attitude and observation skills.

Another interactive way to learn is to search on Internet the necessary information to answer a question or solve a problem. This kind of activity, known as *webquest*, consists of looking for a particular subject which is processed to obtain a given result (Martellini, 2015). Consequently, the didactic materials aren't to be found just in school books but also in internet resources. By using this network research approach it is possible to get in touch with institutions, experts and people outside of the school and then to acquire skills that will come in handy during the whole life.

In order to develop a scientific activity it is a good idea to interact with other students from different countries. This promotes confrontation among several cultures and educational systems and offers significant growth opportunities. These experiences may be realized by using mobile technologies that enhance and support very interesting scientific training projects (Linn, Davis, & Bell, 2004).

Of course, discussing and communicating also means to share.

Sharing videos and pictures is something a youngster would do very often by using its cell phone, and so does a student spontaneously filming experiments, taking pictures during his visit to a museum or his participation to a scientific demonstration. Finally he shares the multimedia contents with other students (Shih, Chang, & Hwang, 2010).

Generally, by using the social network, now also available on mobile phones, it is possible to share videos, websites or scientific papers links, to propose investigations, create albums of pictures: in other words, to improve both the social and cultural dimension.

CONCLUSION

The results of the previously described research activities allowed to define methodological guidelines aiming to support the use of network and mobile technologies in schools, according to a hybrid or mixed point of view. These researches are about the effectiveness of the integration between training processes making use of mobile communication technologies (for instance, distance learning or on line research groups) and face to face training actions based on direct interaction. A proper integration between individual and collaborative study, among activities developed in the classroom and out of the classroom, (for instance, at home), is essential as well. In other words, the use of mobile technologies supporting this kind of communication is not a replacement for traditional training processes, but it provides new opportunities to be effectively combined with one another, according to a wide variety of methodological approaches (Trentin & Bocconi, 2014).

It is generally expected, in educational and family contexts, to promote the dialogue among students, teachers, parents, and the development of WhatsApp groups. Clouds and tablets may be used in schools and in training institutions, together with the already existing virtual environments and "Cloud computing." The number of available applications for the latter has considerably increased in the last years, and it could become a tool devoted to improve creativity and collaboration in educational and social processes.

By doing this it is possible to help each learner to develop his own creativity, to be more aware of his skills, promote personal autonomy, a greater sense of responsibility towards himself, others people, school and social community. The aim is not just to promote the use of the tools but also to increase the learner's production: at the end, he will be able to design and develop specific materials (applet, multimedia papers, etc.) and use them in scientific contexts. This approach will eventually promote his integration into working reality and boost employability.

REFERENCES

Boyd, D. (2014). *It's complicated, The social lives of networked teens*. New Haven: Yale University press.

Duh, H., Tan, G., & Chen, V. (2006). Usability evaluation for mobile device: a comparison of laboratory and field tests. *Proceedings of the 8th International Conference on Human-Computer Interaction with Mobile Devices and Services* (MobileHCI '06), (p. 181–186).

Georgiev, T., Georgieva, E., & Smrikarov, A. (2004). M-Learning - a New Stage of E-Learning. *International Conference on Computer Systems and Technologies* - CompSysTech'2004.

Hofstein, A., & Lunetta, V. (2004). *The laboratory in science education: Foundations for the twenty-first century*. Wiley Periodicals, Inc.

Keengwe, J., & Bhargava, M. (2014). Mobile learning and integration of mobile technologies in education. *Education and Information Technologies*, Volume 19(Issue 4), pp. 737–746.

Kukulska-Hulme, A., & Traxler, J. (2005). *Mobile Learning – A Handbook for educators and trainers*. London and New York: Routledge.

Kukulska-Hulme, A., Sharples, M., Milrad, M., Arnedillo-Sánchez, I., & Vavoula, G. (2009). Innovation in Mobile Learning: A European Perspective. *International Journal of Mobile and Blended Learning,* 1(1), pp. 13–35.

Linn, M., Davis, E., & Bell, P. (2004). *Internet environment for Science Education.* New Jersey: Lawrence Erlbaum Ass. Inc. Publishers.

Martellini, L. (2015). Goerudio: Innovative teaching methodologies in Science education. *Proceedings of 4th International Conference on New Perspectives in Science Education*, (p. 42-45). Florence, Italy.

Sharples, M. (2013). Mobile learning: research, practice and challenges. *Distance Education in China*, 3(5), pp. 5–11.

Shih, J., Chang, C., & Hwang, G. (2010). An Inquiry-based Mobile Learning Approach to Enhancing Social Science Learning Effectiveness. *Educational Technology & Society*, 13 (4), pp. 50–62.

Trentin, G., & Bocconi, S. (2014). The effectiveness of hybrid solutions in higher education: a call for hybrid teaching instructional design. *Educational Technology*, 54(5), pp.12-21.

In: Mobile Learning
Editors: D. René and C. Aubin
ISBN: 978-1-53612-801-7

Chapter 7

BRING YOUR OWN DEVICE (BYOD) IN HIGHER EDUCATION: OPPORTUNITIES AND CHALLENGES

Servel Miller*[*]*, PhD, MSc, PgDip, BSc, FHEA, FRGS, CGeog and Katharine E. Welsh, PhD, FHEA
University of Chester, Chester, UK

ABSTRACT

The recent growth of mobile device ownership provides significant opportunities for universities to utilise these devices to enhance the student experience. Bring Your Own Device (BYOD) allows students to use their own devices in an educational environment. This can provide opportunities such as students feeling more familiar with the technology, enabling personalisation of the devices, and could potentially reduce the cost of keeping up to date with new technology for higher education institutions. However a number of challenges also exist such as how to ensure students without mobile devices feel included and the issue of devices acting as a distraction. This chapter focuses on student's perceptions of BYOD including how motivated they feel to learn and how they use their own devices in the classroom. The findings suggest

[*] Corresponding Author E-mail: s.miller@chester.ac.uk.

that students are willing to use their own devices, they feel more motivated to learn and they feel more connected with teaching overall.

INTRODUCTION

The recent proliferation of mobile digital technologies provides significant opportunities for universities to utilise these devices to enhance the student experience. Whilst mobile digital devices like smartphones are only pocket size, they incorporate computing power and memory capable of running complex software and storing huge amounts of data, that when effectively harnessed could significantly change the way practitioners engage and teach students (Woodcock, Middleton and Nortcliffe, 2012). The National Student Forum Annual Report 2009 challenged universities and lecturers to review their teaching methods to assess whether they are sufficiently taking advantage of new technologies available (e.g., smartphones) (HEFCE, 2010). As highlighted by The Guardian (2015; p1) "There's no doubt that smart technology use can enhance students' experience of university, whether that be keeping in touch with a tutor out of hours or logging on to an online learning environment." In this digital age it is imperative that higher education institutes (HEIs) utilise technology to better engage students and enhance their experience (Gordon, 2014). Technology provides enormous potential for teaching and learning in higher education (Joyce, Boitshwarelo, Phinn, Hill & Kelly, 2014). It is therefore important that HEIs explore creative ways in which to utilise technology, whilst at the same time reducing cost and keeping up to date with technology whilst ensuring that the technology used is fit for purpose and actually playing a role in learning enhancement (Mayes, Morrison, Mellar, Bullen and Martin, 2009). The Bring Your Own Device (BYOD) revolution may address this (Woodcock, Middleton and Nortcliffe, 2012; Johnson, 2012). BYOD is a principle that encourages students to bring their own personal device to educational environments (e.g., school, college, and university) for the purpose of enhancing learning (Alberta Education, 2012). Investment in the most advanced technology can be very expensive but HEIs find the concept of students using devices that they already own to be very attractive (Traxler, 2005). Most (but not all) students currently own smartphone devices. Research by Lerczak (2014) and Medzini (2014) found that 89% of students owned a smartphone and finding a way to utilise their own device (BYOD) in the classroom has both the potential to enhance their learning and reduce the number of devices HEIs need to purchase (Santos, 2013).

Whilst BYOD has the potential to be beneficial to students, staff and HEIs, its impact on student learning has not been fully explored (Jones, Scanlon and Clough, 2012). This paper explores progress in the use of BYOD for learning; student's attitudes towards the perceived benefits and limitations of BYOD; if and how BYOD can contribute to learning and/or motivating students to learn; the challenges of implementing BYOD; and whether or not BYOD could potentially marginalise sections of the student population. By answering these questions it may be possible to better develop an effective BYOD strategy and infrastructure for HEIs (Jones, Scanlon and Clough, 2013).

BYOD: Opportunity or Threat?

Generally, university staff and students are not fully aware of the potential that mobile devices offer to support learning (Woodcock et al., 2012). Whilst students generally do not install applications (apps) on their own devices for the purposes of 'in class' learning, they are generally very open to the use of their own devices to support their education (HEFCE, 2010). As highlighted by Woodcock et al. (2012, p13) "once students begin using their smartphones for learning they begin to appreciate the benefits and the further possibilities that exist within their current frame of reference." Use of mobile technologies has the potential to benefit students immensely but may also bring its own share of challenges. These need to be explored and understood to effectively implement a BYOD strategy in HEIs (Cochrane and Bateman, 2010).

Advantages of BYOD

There are many advantages of using BYOD schemes within learning environments. BYOD gives students the ability to work more flexibly, with easier access to university Wi-Fi (compared to traditional desktop computers), thereby maximising their engagement and making productive use of their time (BT iNet, 2012). At the University of South Florida for example, students and faculty benefit from continuous, secure access to online applications for interactive teaching and learning (Cisco, 2012). Students can also use their mobile devices to respond to multiple-choice questions during lectures, allowing professors to collect active feedback (Cisco, 2012). Research undertaken by JISC (2012) found that there were many benefits of using BYOD including:

- Students being able to use their own devices rather than imposing technology upon them
- Students being given the opportunity to personalise their device, which in turn allows them to access and engage with their learning in ways which meet their individual needs
- Greater engagement with the topic area
- Increased student satisfaction, retention and widened participation.

A study by Conole et al. (2008) also found students placed greater value on technologies that they had discovered for themselves which may in turn increase student motivation. However, Linsey et al. (2010) found that often students need to be prompted to use their own devices for their learning with specific activities. This puts the responsibility on the educator to understand the educational value of BYOD and to be aware of how to integrate BYOD into classrooms (Fang, 2009). Welsh & France (2012) outline how students "typically view their smartphone as a basic communication tool and often do not exploit the rich potential that these devices hold for learning" (p48) yet Conole et al. (2008) found that students use their own devices extensively outside of the classroom for learning.

With a greater degree of personal ownership and personal finance involved, students may be more likely to take the time to maintain their devices and as such they are always readily functional and sufficiently 'up-to-date' to address their needs (Wood, 2012; Afreen, 2014). In some instances, the student owned devices are often of higher quality than those offered by the institution (Medzini et al., 2014). This is turn may benefit universities by reducing the cost of updating and purchasing new devices (Santos, 2013). The device brought by students is familiar to them (Ballagas et al., 2004) and reduces the time needed to train students in how to use that piece of technology. As students already own such devices, there is the opportunity to focus more on the learning potential rather than technological difficulties (Bedall-Hill, 2011). They can also become more familiar with the software or apps as they have more time to use them at home as well as in the classroom (Pfoutz, 2012). Familiarity with the device allows students to engage more and collaborate with both staff and fellow students (EDTECH, 2012; p2). Furthermore, students with Specific Learning Difficulties (SpLD) and visual impairment can use their own disability software (Grussendorf, 2013) and change the text size, format and colour to suit their needs.

Challenges of BYOD

Whilst BYOD has many benefits there may some barriers to its implementation. For example whilst the initial costs of BYOD are relatively small, the long term support required at institutional level is greater because of the additional pressures on networks and support associated with different models of mobile devices (Afreen, 2014). This may be the biggest challenge facing the effective use of BYOD particularly for HEIs that have not been proactive in updating their existing network infrastructure. The variety of equipment owned by students, with different operating systems, may result in compatibility problems with HEI's network system (Kobus, Rietveld, and van Ommeren, 2012), infrastructure and online resources. BYOD schemes must to be able to operate within the confines of the devices owned by students, the alternative being increased investments made by HEIs in suitable mobile devices for the classroom (Educause, 2012). In addition, externally acquired devices may need to undergo safety checks to ensure they meet existing health and safety standards before they can be used and/or sanctioned to be utilised on university networks (JISC, 2012).

Implementing BYOD could also potentially have a negative impact on learning. Students bringing their own devices into the classroom has the potential to serve as a distraction rather than a tool to enhance learning. As highlighted by The Guardian (2014; p1) "Students' own devices are likely to include all the applications they use on a regular basis. This cannot be controlled like it can be with computers provided by the institution." It may be very difficult to maintain students' attention when they may be preoccupied with their various apps and/or engaged with social networking which may impact on their learning (Miller et al, 2012, p53). Students on average *spend around six minutes listening before switching to technology (Rosen, 2013).* This distraction could impact on their long term ability to concentrate on tasks. At a university level where deep analytical thinking is critical to student learning this could be problematic and a serious barrier of effectively utilising BYOD (Rosen, 2013).

BYOD and Inclusivity

A major challenge of implementing an effective BYOD strategy is to ensure there is no inequality between those that have the means to purchase and use their own device and those who do not. Stager (2011) postulated that the BYOD revolution could 'institutionalize' inequality of how students access technology. Students from less privileged backgrounds are far more likely to be marginalised. However, recent studies have shown that a high percentage of students already own a mobile device (Dixit *et al.*, 2011) and research by JISC (2010) suggested that by 2015, 80% of students would be accessing the internet via a mobile device. This percentage will likely increase in the future due to advances in technology and devices becoming more affordable (Medzini et al., 2014), therefore inclusivity due to lack of device could become less of a problem in the future. Nevertheless, it is unlikely that 100% of students will have a smartphone, particularly with sufficient capabilities to enhance learning, within the next five years. Therefore, in the short term, the inclusivity aspect of BYOD still must be addressed. The current response to potential inequality is to allow students to bring their own devices and also supply those who do not have a suitable device with an institution-owned device (Vasant, 2015). Whilst they may lose their ability to personalise the device and be less familiar with the institutional device than they would with their own device, the student would still have the capability to carry out the required tasks in the classroom. In order to address the additional inequality between devices, BYOD exercises should be designed to ensure that they are "device neutral" (Alokaily, 2013) or make use of multi-platform software (e.g., Evernote, Dropbox) (Shepman et al., 2012) to ensure inclusivity where possible.

Despite the discussion surrounding BYOD in terms of lack of inclusivity, there are opportunities to be gained from potentially making use of BYOD for students with SpLD such as dyslexia or students who are visually impaired. Studies by Swan et al. (2005) suggest that teachers perceived there to be a smaller gap in academic achievement between typical students and those with additional needs when mobile devices are used and raise the idea that students personalise the devices to their own needs. This benefit extends to students without additional needs and conforms to the ideas of familiarity outlined earlier. It could be said therefore that rather than excluding or marginalising students, BYOD has the potential to make classrooms more inclusive.

BYOD and Learning: Opportunities

Traditionally many educators have viewed mobile devices or implementation of mobile technologies negatively (Hernandez, 2013;

Barry et al., 2015) assuming they are a distractor which may in turn leave students feeling as though they are not able to make use of their own device for learning. Students spend much of their time outside of the classroom using their mobile devices yet they are often banned from classrooms in favour of pen and paper (Norris & Soloway (2010). As Vanwelsenaers (2012) states "We claim to be teaching 21st -century skills but are using 18th-century tools" (p.8). As highlighted by(Kobus (2012) and Gordon (2014) it is difficult to argue with the benefits of BYOD, especially when the strategy endorses and encapsulates the notion of flexible working However, the impact on learning and whether it motivates students to learn needs to be explored. BT iNet (2012) suggests that two thirds of organisations have no BYOD strategies in place. As outlined here, BYOD seems like an attractive prospect for HEIs, however relatively little research has been conducted into student perceptions of BYOD in a classroom setting. Some of the limited studies conducted into student perceptions of using their own devices in a classroom setting suggest that students often perceive that their devices are being "invaded by academic materials" (Jarvis & Dickie, 2010, p.77) and "an infringement of norms of behaviour between teacher and learner" (Knight, 2006, p.21) when they have been asked to use their personal devices in an academic setting. Yet Bedall-Hill (2011) found that students used their devices for both educational and social purposes which supports the findings of Welsh et al. (submitted) who found that 75% students surveyed would be happy to use their own device for fieldwork. This perhaps suggests a shift in student perception when compared to some of the earlier studies (e.g., Knight, 2006; Jarvis & Dickie, 2010; Woodcock, Middleton and Nortcliffe, 2012) that found students who own smartphones are largely unaware of their potential to support learning and, in general, do not install smartphone applications for educational purposes. This shift in student thinking is possibly due to a combination of these technologies being almost ubiquitous across the population, a reduction in cost of devices, devices not viewed as being fragile and difficult to replace if broken, or simply that HEIs may be doing more to promote the idea of BYOD. This needs to be explored and is the focus of this paper.

Motivation

Practitioners are eager to improve students' motivation levels and have, in the past, looked to using new technologies to boost the motivation levels of students (Rau et al., 2008). McNeill et al. (2011) outline how the current generation of students "have a low tolerance for lectures [...] and where technology can be used to enhance engagement this should be encouraged" (p6). Technology that encourages more interaction between students and tutors and has a fun/entertainment component (e.g., video clip/educational game) has proven to motivate students to learn but it important to ensure that such tools are not too distractive (Miller, 2012; Martin, McGill, & Sudweeks, 2013). Studies have found that smartphones can be used to "create new and exciting learning experiences" (Medzini et al., 2014 (p15) and therefore increase a student's motivation to learn (Swan et al., 2005; Jones et al., 2006). As such, this paper will explore whether BYOD as a tool motivates students to learn or engage more with their studies or if it is just another technology that serves to distract students from the main learning task in lectures.

METHODOLOGY

For this research, level 4 (first year) students on a Geography undergraduate programme participated in a survey. This provided an opportunity to obtain a large enough sample group which was representative of the student body in the department. In total, there were 91 student participants out of 140 students in the class. As part of this research, students were encouraged to bring their mobile devices into class to be used as part of a number of teaching sessions. The range of activities students utilised their own devices for included: to access lecture slides; to access the internet to read a journal article which had been discussed in class (see Figure 1); to send responses to tutors (e.g., to questions posed in class); to seek clarification and/or send comments (e.g., via Poll Everywhere) to tutors, which were projected onto a screen; to aid student presentations in class; to capture output from group work activities (e.g., a poster made in class or web diagram using the camera function-see Figure 2) and generally to share information between groups.

Figure 1. Students using their own device to read a journal article in preparation for class discussion.

Figure 2. Students capturing details from a poster made and discussed in class.

Development of the Survey

The survey was designed to explore key themes as outlined in the research objectives (see Section 1). A range of question types were used including closed multiple choice and 'Likert' scale questions for a "quick-look" (Bird, 2009), and open-ended questions which enable participants to share their responses with greater depth (McGuirk & O'Neill, 2005). Thematic analysis (Guest, MacQueen & Namey, 2011) was undertaken with the qualitative data available following the approach demonstrated by Braun and Clarke (2006).

The CDW (2012) found that technology is slowly being adopted into college and university curricula. Notably, 31% of students used technology as a learning tool while in class in 2011 (up from 19% in 2010). In 2013 (Bradford Networks, 2013) this was up to 75%. This research reports the use of a 'Pervasive BYOD strategy' in some HE institutions to foster this trend. First of all the questionnaire was designed to ascertain student's access and progress in the use of mobile technologies for the purpose of their learning. As such, questions such as "which of the following portable/mobile electronic devices do you own?" and "which of the following academic related tasks do you use your own portable/mobile electronic device to do?" were included.

The second objective of this research was to evaluate students' changing perceptions and attitudes to using their own devices in class. Whilst it was important to evaluate and quantify the percentage of students willing to use their devices for learning, the research also wanted to gather students' views in a qualitative manner to identify the benefits and limitations of BYOD for students. As such, questions such as: "Do you think it is appropriate for students to use their own device in class sessions? If so, why?" and 'What are some of the benefits and limitations of using your own device in class sessions' were included to evaluate students attitude to the use of BYOD.

The third theme explored was how using their devices impact on their learning and whether the device motivates them to engage more with what is being taught. In order to explore this, students were ask for example; to rank a range of question (on scale from 1-5, with 1 being not all and 5 being very much) including; 'Makes me feel more connected to tutors', 'Makes me feel more connected to my programme of study' and 'Make me want to learn more'. At the end of the questionnaire, students were asked to provide qualitative responses to support their rankings.

The final theme explored was inclusivity. As part of the research we wanted to explore if some students felt uncomfortable and/or could not afford the technology and therefore would feel excluded. The key questions asked was therefore; '*On a scale of 1 to 5, with 1 being very excluded and 5 being very included, how do you feel when you are asked to use your own device in classroom settings and to explain your answer'*. As before students were asked to provide qualitative responses to support their responses.

Administering of the Survey

The surveys were made available online through the Survey Monkey website. Once the students had completed the class sessions where they were able to use their own devices, a link to the website was provided both in class sessions and also emailed to the students. In this case the target sample population was 140 students. Students were only allowed complete online survey once. The online survey was available for up to two months after the final class activities (see Section 3) to ensure that students had sufficient time to reflect on their experiences which would hopefully better inform their views. In total 91 questionnaires were completed, which represented 65% of the target population.

Results and Discussion

Of the 91 students who provided survey responses, 92% (n=91) were between 18 and 21 with the other 8% being mature students (22+). When students were asked to rank their IT skills and understanding using a Likert scale (1 being very limited, 5 being excellent), most students ranked their skills at 4 or 5 (80%), with approximately 10% ranking their skills as excellent. Less than 2% ranked their skill and knowledge as poor. These results indicate that students are generally capable users of IT technology, regardless of whether they are typical under 21 undergraduate students or mature students.

Student Perceptions of BYOD

Students were asked to describe what they thought of BYOD in three words. The words provided were used to create a word cloud. The words that are more pronounced in size indicate a higher frequency of answers for giving that word (Figure 3). When the word cloud was visually analysed, it is evident that the most prominent words used by students were; Interactive, Fun, Modern, Helpful and Easy. This suggests an overall positive view of BYOD as a concept by students. The fact that students highlighted "interactive" as one of the most prominent words based on their experience is a positive sign. Improving interactivity in lectures or teaching sessions can enhance student learning as "students who contribute enthusiastically in lectures retain information for longer than if they simply see or hear it" (Karanicolas et al., 2011; p666). In addition, increased levels of student motivation are also thought to be linked to interactivity and active participation in lectures (Machemer and Crawford, 2007; Miller and France, 2013). It could therefore be said that BYOD has the potential to improve student motivation levels due to enhanced interactivity in the lectures. Students perceive using their own devices to be "modern" which may improve their overall satisfaction with the module or course, particularly if they feel as though they are using "cutting edge" technology which students now expect to use in degree programmes (HEFCE, 2010). It is encouraging to find that students think using their own device is "easy" as for too many years, technology has quite often been viewed as "difficult" or "technical" which often proves to be a major barrier by students who are less technically able (Welsh et al., 2013). Students perceiving BYOD positively is a further indication that this technology is being used to enhance learning rather than being a distraction (Rosen, 2012).

Figure 3. Word cloud showing the most popular words gathered from students who were asked to describe BYOD. Larger words indicate more frequent responses.

How Beneficial Is the Use of IT Technology in Class Sessions?

When students were asked 'how beneficial is the use of IT in class sessions?' overwhelmingly their responses were positive. Overall, 95% of students (n=89) believed that IT use in the classroom was beneficial to them to varying degrees. 11% thought it was extremely beneficial, 37% believed it was very beneficial and 47% believed it to be beneficial. Only 5% thought that the use of IT in class sessions was marginally beneficial and no students thought that IT was not beneficial. The main reason cited by respondents as to why it is beneficial, was the ability for them to follow the lecture at their own pace. This was exemplified by student comments such as "I can look at the resources whenever I want and can take my time to read through them at my own pace" and "I use my device to look at the PowerPoint that is being shown in class, in case we don't get all the notes for that slide wrote [sic] down in time." One of the students also stated that they found using their own devices beneficial as it helped them to "engage with the lecture material, contribute to discussion and gain a greater understanding of concepts being explained." Furthermore, students found using their device for additional research during the lecture session to be beneficial as highlighted by one student who said, "I can find out definitions for certain words or phrases if I do not want to ask." Whilst many students outlined the benefits of using their own devices, several students also commented for example “devices may be a distraction” and “use of social media may be a distraction.” The potential for BYOD to be a distraction has been highlighted by Rosen (2012) which has become evident from this research. Whilst this did not appear to be a major worry by the students, tutors need to be aware of this pitfall, implement mechanisms to address this. Preventing students from using for example social media whilst utilising BYOD is not enforceable but awareness can be raised amongst students regarding its distraction potential and how it may impact negatively on their learning.

From the survey responses, students appear to engage with technology and recognise the benefits. As highlighted by the word cloud (Figure 3) where students indicated that the use of BYOD was ‘helpful’ and the 91% of students who perceived IT as being beneficial, it could be said that technology is not a barrier in itself and should not serve to disengage some student groups. There may be factors that contribute to BYOD being a barrier, some of which are explored later in this paper, but a student’s lack of familiarity with technology itself does not appear to be one of them and as such, there is merit in utilizing BYOD more in the classroom.

Appropriateness of Using Own Devices in Classrooms

A key objective of this research was to evaluate students’ attitudes to using their own devices. As such, they were asked whether they felt it was appropriate to use their own device in the classroom educational setting. From the responses provided, 95% stated it was appropriate to use their own mobile devices in the classroom/lecture theatre with only 5% said they felt it inappropriate.

This research wanted to go beyond quantifying student’s willingness to use their own devices and explore their views as to why they thought it was appropriate. When the qualitative responses were analysed, firstly, the most popular reason cited was linked to engagement and interest. This is reflected in student comments such as “helps engage students through interactive learning" and "keeps people interested." This was also supported by results highlighted by Figure 3, when students were asked to provide three words to sum up their experience and their top response was 'interactivity'. BYOD provides this interactivity which students have recognised and which, as demonstrated in this research, they highly value.

Secondly, a key theme that has emerged is students being more receptive to BYOD because of the ease of use. A typical statement that summarises this idea was that using their own device was "easier as you are used to how it works." This point highlights the importance of familiarity, where students understand their own personal devices and are more confident in using them. As one student indicated it is "quicker" and "easier than handing out iPads to everyone at every lecture," thereby saving a small amount of time during lectures.

Of the 5% who did not think using their own device was appropriate, the main reasons cited were that "some do not have smartphones" and that technology is "often pointless and a waste of time." Whilst the overwhelming majority of students believe it is appropriate to use their own devices in class, there is still a small number which believe it is not. This needs to be taken in consideration when developing a BYOD strategy. Questions still need to be asked about BYOD including, how to engage these students who may be resistant to the idea? What if they do not have a device? How to ensure they are able to participate and engage with activities that require a mobile device? One solution is to have a number of devices (e.g., tablets), readily available that students may simply collect and use during the session. This will require some investment by universities in these resources but BYOD provides an opportunity to reduce this investment yet investment is still needed for BYOD strategy to work effectively.

The findings of this research show conclusively that the overwhelming majority (95%) of students believe it is appropriate to use technology in a classroom environment. As with previous research (see Bradford Networks, 2013; CDW, 2012), there has been a steady increase in students willingness and attitude to using their own devices. Again this is demonstrated within the findings presented with 95% willing to use

BYOD, which compared to 75% in 2013 (Bradford Networks, 2013). This demonstrates an opportunity for HEIs to exploit and use BYOD to their benefit.

BYOD and Inclusivity

Woodcock et al. (2012) research indicated that just below 70% of students had access to personal mobile devices. JISC (2010) suggested that by 2015, 80% of students would be accessing the internet via their personal mobile device. By this projection, approximately 20% of students could be marginalised by a BYOD strategy if they do not have their own mobile device to utilise for in-class activities. Whilst this may only represent a small number of students in each class, across the UK, 20% of students without devices equates to approximately 460’000 (HEFCE, 2014), thereby indicating the level of investment still required from HEIs into institutional devices. A key focus of this research was to identify whether there are a sufficient number of students who have access to mobile technology that they can utilise in class. The results indicated that over 92% of students have access to/own a smartphone and 100% of them own or have access to a laptop. However, what was more notable was that no students indicated that they do not possess or have access to a portable device that could be used in class (see Table 1). The research findings therefore suggest that access to portable devices is not a significant barrier to BYOD and should not affect the implementation of such strategies in HEIs. However, there is still a small proportion of the student population that may still be marginalised and HEIs must put in place mechanisms to reduce this possibility. One solution is for HEIs to purchase a number of mobile devices which students can readily and discreetly have access to during class sessions. Institutional schemes are in place in some HEIs which allow students to book a device (e.g., laptop) from the library in advance which would go some way to reduce any feelings of embarrassment which may arise from walking to the front of a class to pick up an institutional device, thereby making them feel marginalised.

Table 1. Mobile devices that students have access to and/or owned

Smart Phone	92%
Mobile Phone (not a smart phone)	15%
Blu-ray Player portable	0.%
Laptop Computer	100%
Tablet computer (e.g., iPad, Samsung Galaxy)	38%
Digital Camera	69%
E-book Reader (e.g., Kindle, Nook)	15%
GPS Device (e.g., Garmin)	30%
MP3 Player (e.g., iPod shuffle)	62%
I don't own any of these electronic devices	0.00%

BYOD and Learning

One of the key research questions was to explore whether students believe using their own devices has an impact on their learning. 53% (n=91) of students responses indicated that using the devices made them want to learn more (see Table 2). A key theme about how BYOD encouraged them to learn more was the level of independence and perceived autonomy it provided them. Sweeney (2012) alluded to this independence when explaining that BYOD changes the entire teaching model from one where teachers control what students deliver (e.g., "use PowerPoint to do X") to one where the student controls their own delivery (e.g., the teacher says "demonstrate your knowledge of X.") in their own style and creativity. This level of independence was highlighted by students who said for example; "allows independent study"; "could do more research around topics brought up in class" and "we don't get treated like high school kids, it just helps understand some modules better." The positive manner in which this affected their learning was for example the ability to better follow and engage with lectures as well as using additional resources to help them learn more. For example students commented "Yeah, more engaging and use of more resources"; "It just helps show what other opinions are within the room" and "Power points created by lecturers make lectures easier to understand. Documents on the module space make revision easier and lectures easier to engage with. Using personal devices enables a greater level of student contribution." From the responses for the survey, it is clear that the use of BYOD encourages a more thorough connection to the module (see Table 2) with students suggesting it allows more flexibility in accessing learning resources, which enhances their learning process. This is reflected in such comments as; "It also speeds up the process and makes learning more interactive and exciting. I can look at the resources whenever I want and can take my time to read through them at my own pace" and "It helps me build a variety of skills that are useful in aiding and supplementing other module learning." As highlighted by Figure 4, when students were asked how motivated they felt to participate in class sessions when using their own devices, the majority of students (73%, n=91) feel motivated to participate in class sessions when using their own device. A further 19% feel neither motivated or unmotivated and just 9% (n=91) felt as though they were not more motivated to participate in class sessions when they used their own device. Though some literature points to mobile devices being a distraction in lectures (e.g., Geist, 2011) and there being some resistance to embrace BYOD in HE (Johnson et al., 2011), Scornavacca et al. (2009) found that generally the devices can have a positive impact which supports the findings presented within this paper.

Knight's (2006, p.21) assertion that students view the use of BYOD as "an infringement of norms of behaviour between teacher and learner" seems not to be borne out by this research. As highlighted by Table 2, over 60% of students indicated that BYOD enables a positive connection with tutors, with only 15% feeling the opposite and approximately 23% felt neutral about this connection. The days of students feeling as though using personal devices for academic purposes is an invasion of their "personal space" (Jarvis & Dickie, 2010; p77) appears to be subsiding and this is evidently not such as issue in the current era. Though the findings presented here suggest that the majority of students found BYOD motivated them to participate in the lecture more, other studies have demonstrated that motivation often declines after the "novelty effect" of the first few sessions has worn off (Veletsianos and Navarrete, 2012; Swan et al. 2005). Whilst this research spans an academic year and highlights the motivation that using BYOD provides, it does recognise the need for longitudinal research over a longer time period (for example through three years of study) to ascertain this. The research into BYOD over longer time scales is ongoing and the findings will be presented in a later paper.

Table 2. Responses exploring the BYOD and it relation to 'connection' with module, tutor, department, course and fellow student

Rankings (In relation to using your own device in classes, could you please rank the statements below, with 1 being not all and 5 being very much)	**1 –**	**2 –**	**3 –**	**4 –**	**5 –**
Makes me feel more connected to the module	7.69%	30.77%	7.69%	23.08%	30.77%
Makes me feel more connected to tutors	0.00%	15.38%	23.08%	53.85%	7.69%
Makes me feel more connected to my department	0.00%	0.00%	61.54%	38.46%	0.00%
Makes me feel more connected to my programme of study	0.00%	15.38%	23.08%	38.46%	23.08%
Make me want to learn more	0.00%	15.38%	30.77%	23.08%	30.77%
Make feel connected more to fellow students	0.00%	46.15%	38.46%	7.69%	7.69%

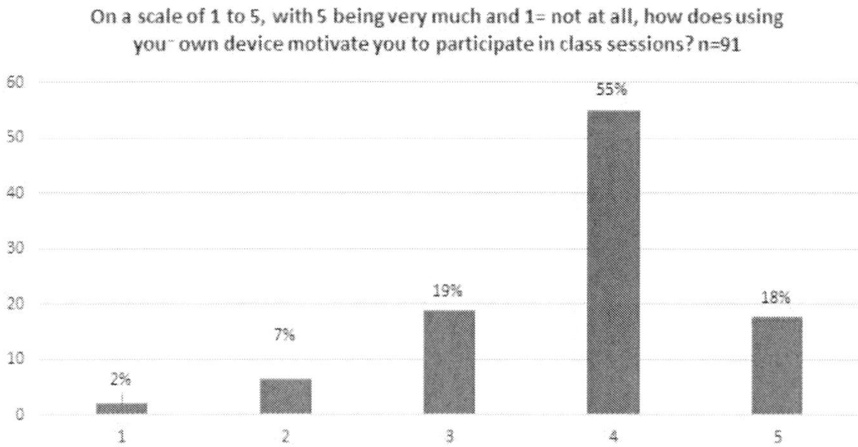

Figure 4. Student perceptions of how motivated they feel to learn when using their own device.

Challenges

Whilst students highlighted that BYOD provides them with an opportunity to be a more independent learner and motivates them to learn more in class, they also highlighted some challenges of BYOD. The main challenge is access to the university network and students indicated the lack of access in some classrooms and the slow network connection. This was highlighted by a students who said "Wi-fi is not available in all areas." Lack of internet connection could disrupt learning activities which has the potential to reduce the impact of BYOD. The lack of sufficient wireless connectivity in the classroom (e.g., not available in some sections of the room) could also disenfranchise some students, as it prevents them from participating fully in class. The students who are unable to access or complete the task may become disengaged and disruptive, which in turn can affect the engagement levels of rest of the group. Students highlighted that BYOD itself might be disruptive: "keeps easily distracted students occupied and interested in class work but can be distracted by social media" and "makes an academic use for phone but people could abuse it by using phone for other reasons" which thus introduces a new challenge to the classroom. It is therefore imperative that for a BYOD strategy to be effective and implemented successfully across HEIs, connection issues are addressed otherwise gains made by using BYOD may be easily eroded and their use may become counterproductive.

CONCLUSION

The research presented within this paper highlights that more students have access to mobile devices compared to a couple of years ago and more significantly are willing to use these for academic purposes. Students (both young and mature) are very comfortable with IT and unlike a few years ago where this could be a barrier, this seems not to be the case. From teaching and learning perspective students perceive BYOD to be beneficial to them and as such, it is recommended that BYOD be implemented more widely. The findings show that BYOD can help students to engage and connect with many aspects of being in a classroom, such as the content, the tutor and the department. Many students found BYOD saved timed during lectures and provided them with the opportunity to be a more independent learner. This is due in large part to their familiarity with the technology, where they are comfortable with the devices they own and their capability. There is still a limited number of students does not own mobile devices which may results in them being marginalised. BYOD is rolled out more widely, it will be necessary for institutions to provide some mobile devices for students to supplement those that do not own one. The findings suggest that whilst BYOD has the potential to reduce institutional costs of purchasing up-to-date technologies, it cannot eliminate the cost completely, at least for the medium term. One of the difficulties that may be faced by practitioners is the nature of the exercise that they ask a class to participate in and the need to ensure compatabilities with multiple devices (e.g., an exercise is not specific to an app which is only available on certain devices or platforms). If this is not done, there is the risk of disfranchising some students from the task at hand, which may prove counterproductive and dilute the value of BYOD. Successful implementation of BYOD for specific learning activities should be carefully designed to ensure multi-platform software can be used. Whilst this study has indicated the use of BYOD can improve motivation, longer-terms studies are necessary to identify whether this is a "novelty" effect that wears off with time. BYOD has been shown be beneficial to students' learning but HEIs need to implement the appropriate infrastructure (e.g., access and speed) to ensure they reap the rewards of this emerging 'technology'.

REFERENCES

Alberta Education, (2012) *Bring your own device: A guide for schools.* Alberta Education. Retrieved from http://education.alberta.ca/admin/technology/research.aspx.

Ansaldi, H. (2013). *Addressing the Challenges of the 'Bring Your Own Device' Opportunity.*

Afreen, R. (2014). Bring Your Own Device (BYOD) in Higher Education: Opportunities and Challenges. *International Journal of Emerging Trends & Technology in Computer Science,* 3(1).

Al-Okaily, R. (2013). Mobile learning and BYOD: implementations in an intensive English program. *Learning and Teaching in Higher Education: Gulf Perspectives, 10*(2).

Ballagas, R., Rohs, M., Sheridan, J. G., & Borchers, J. (2004, September). Byod: Bring your own device. In *Proceedings of the Workshop on Ubiquitous Display Environments, Ubicomp* (Vol. 2004).

Barry, S., Murphy, K., & Drew, S. (2015). From deconstructive misalignment to constructive alignment: Exploring student uses of mobile technologies in university classrooms. *Computers & Education, 81*, 202-210.

Bedall-Hill, N. (2011). Postgraduates, field trips and mobile devices. In J. Traxler & J. Wishart (Eds.), *Making mobile learning work: Case studies of practice* (pp. 18–22). ESCalate HEA Subject Centre for

Education. Retrieved September 3, 2011, from http://escalate.ac.uk/8250.

Bird, D.K. (2009). The use of questionnaires for acquiring information on public perception of natural hazards and risk mitigation – a review of current knowledge and practice. *Natural Hazard Earth System Science,* 9, 1307–1325.

Bradford Networks (2013) Education Institutions at Forefront of BYOD, *Infosecurity magazine*, 22 May, 2013[online]. Available: http://www.infosecuritymagazine.com/view/32549/education-institutions-atforefront-of-byod-/.

Braun, V. and Clarke, V. (2006) Using thematic analysis in psychology. *Qualitative research in Psychology,* 3 (2) p. 77-101.

BT iNET (2012) *BT iNet helps Brunel become more attractive to students through Bring Your Own Device.* Retrieved 5th May 2015 from www.insight.bt.com/downloads/.../inet/brunel_university_case_study.pdf.

Cochrane, T. and Bateman, R. (2010) Smartphones give you wings: pedagogical affordances of mobile Web 2.0. *Australasian Journal of Educational Technology*, 26 (1), 1-14.

Conole, G., de Laat, M., Dillon, T., & Darby, J. (2008). Disruptive technologies, pedagogical innovation. What's new: Findings from an in-depth study of students' use and perception of technology. *Computers & Education,* 50(2), 511-524.

CISCO (2012) *BYOD in Education.* Available from: http://www.cisco.com/web/strategy/docs/education/46096_byod_ed_aag.pdf.

CDW, 2012. "Bring Your Own Device: Adapting to the flood of personal mobile computing devices accessing campus networks." *EdTech: Focus on Higher Education*. February.

Dixit, S., Ojampera, T., Nee, R. and Prasad, R. (2011) Introduction to globalization of mobile and wireless communications: today and in 2020. In: Prasa, R. Dixit, S.van Nee, R. and Ojanpera, T. (eds) *Globalization of Mobile and Wireless Communications Signals and Communication Technology*, Springer Science and Business Media, 2011, 1-8.

Draffan, E. A., Evans, D. G., & Blenkhorn, P. (2007). Use of assistive technology by students with dyslexia in post-secondary education. *Disability & Rehabilitation: Assistive Technology*, *2*(2), 105-116.

Educause. (2012). *Bring Your Own Device* (BYOD).

Fang, B. (2009). From distraction to engagement wireless devices in the classroom. *EDUCAUSE Quarterly*, 32. (4). Retrieved from http://www.educause.edu/ero/article/distraction-engagement-wireless-devices-classroom.

Guardian (2015) *Why UK higher education institutions should invest in technology*. Available form http://www.theguardian.com/higher-education-network/blog/2011/apr/18/higher-education-investing-in-technology.

Guardian (2014) *Students: bring your own technology to uni.* Available from: http://www.theguardian.com/education/2014/apr/11/students-bring-tech-device-uni.

Gordon, N, (2014) *Flexible Pedagogies: technology-enhanced learning.* The Higher Edauction Academy. Available from: https://www.heacademy.ac.uk/sites/default/files/resources/TEL_report_0.pdf.

Grussendorf, S. (2013). *Device ownership, 'BYOD' & social media for learning.* Chicago.

Geist, E. (2011). The game changer: Using iPads in college teacher education classes. *College Student Journal,* 45(4), 758.

Guest, G., MacQueen, K. M., & Namey, E. E. (2011). *Applied thematic analysis.* Sage.

Hernandez, E.L. (2013) *Mobile learning: education with the help of your cell phone.* Retrieved from http://assett.colorado.edu/mobile-learning-education-with-the-help-of-your-cell-phone/ on 23/04/15.

HEFCE (2010) *Student perspectives on technology – demand, perceptions and training needs: Report to HEFCE by NUS.* Last accessed 20 September 2011. Online at: http://www.hefce.ac.uk/pubs/rdreports/2010/rd18_10/.

Jarvis, C. and Dickie, J. (2010) 'Podcasts in support of experiential field learning', *Journal of Geography in Higher Education,* 34, pp. 173–86.

JISC (2012) *Bring Your Own Device (BYOD*) Available from: http://www.jisc.ac.uk/blog/bring-your-own-device-byod-23-nov-2012.

Jones, A.C., Scanlon, E. and Clough, G., (2012) Mobile learning: Two case studies of supporting inquiry learning in informal and semiformal settings. *Computers & Education* 61 (2013) 21–32.

Johnson, L., Smith, R., Willis, H., Levine, A. and Haywood, K. (2011) *The 2011 Horizon Report.* Austin, TX: The New Media Consortium.

Johnson, D. (2012) On Board with BYOD *Educational Leadership,* October 2012.

Joyce, Boitshwarelo, Phinn, Hill & Kelly, 2014.

JISC blog. (2012). Bring Your Own Device (BYOD).

Kobus, P. Rietveld, J.N. van Ommeren (2013) Ownership versus on-campus use of mobile IT devices by university students. *Computers & Education*, 68 (0) (2013), pp. 29–41.

Knight, J. (2006) 'Investigating geography undergraduates attitudes to teaching, learning and technology', *Planet,* 16, pp. 19–21.

Jones, A. C., Scanlon, E., & Clough, G. (2013). Mobile learning: Two case studies of supporting inquiry learning in informal and semiformal settings. *Computers & Education, 61*, 21-32.

Lerczak, A. (2014) Student Perceptions of Using Mobile Technologies in Higher Education Fieldwork. Unpublished Masters Dissertation.

Linsey, T., Hall, R., Ooms, A., Taylor, P., Andrew, M., O'Brien, J., Field, K. and Downward, S. (2010). Transforming Curriculum Delivery through Technology. *JISC Final Report.*

Martin, R., McGill, T., & Sudweeks, F. (2013). Learning anywhere, anytime: student motivators for m-learning. *Journal of Information Technology Education: Research*, *12*(1), 51-67.

Mayes, T., Morrison, D. Mellar, H., Bullen, P and Martin, O., (2009) *Transforming higher education through technology-enhanced learning* The Higher Education Academy Innovation Way York Science Park Heslington.

McGuirk, P. M. and O'Neill, P.: Using Questionnaires in Qualitative Human Geography, in: *Qualitative Research Methods in Human Geography*, edited by: Hay, I., Oxford University Press, Australia, pp. 147–162, 2005.

McNeill, M., Ming Diao, M., & Gosper, M. (2011). Student uses of technology in learning: two lenses. *Interactive Technology and Smart Education*, *8*(1), 5-17.

Medzini, A., Meishar-Tal, H., & Sneh, Y. (2015). Use of mobile technologies as support tools for geography field trips. *International Research in Geographical and Environmental Education*, *24*(1), 13-23.

Miller, K. Voas, J. and Hurlburt, G. (2012). BYOD: Security and Privacy Considerations. IT Professional 14 (5). pp. 53-55. *Norris & Solway* (2010).

Rau, P., Gao, Q., & Wu, L. (2008). Using Mobile Communication Technology in High School Education: Motivation, Pressure, and Learning Performance. *Computers & Education,* 50, 1-22.

Rinehart, D. (2012). *Students Using Mobile Phones in the Classroom: Can the Phones Increase Content Learning.* Retrieved from ProQuest Digital Dissertations. (UMI 1517773) Rosen, L (2013) *Distracted? How hyperstimulation is making you smarter. Wired issue 10.13.*

Santos, I. M. (2013, October). Use of students' personal mobile devices in the classroom: Overview of key challenges. In *World Conference on E-Learning in Corporate, Government, Healthcare, and Higher Education* (Vol. 2013, No. 1, pp. 1585-1590).

Schepman, A., Rodway, P., Beattie, C., Lambert, J. (2012). An observational study of undergraduate students' adoption of (mobile) note-taking software. *Computers in Human Behavior, 28(2), 308 – 317.*

Scornavacca, E., Huff, S., & Marshall, S. (2009). Mobile phones in the classroom: if you can't beat them, join them. *Communications of the ACM,* 52(4), 142-146.

Stager, G. (2011, October 8). *BYOD: Worst idea of the 21st century*? [blog post] Retrieved from *Stager-to-Go* at http://stager.tv/blog/?p=2397.

Sweeney, J. (2012). *BYOD in Education: A report for Australia and New Zealand.*

Swan, K., Hooft, M. V. T., Kratcoski, A., & Unger, D. (2005). Uses and effects of mobile computing devices in K–8 classrooms. *Journal of Research on Technology in Education*, *38*(1), 99-112.

Traxler, J., & Kukulska-Hulme, A. (2005). *Evaluating mobile learning: Reflections on current practice.*

Vasant, S. (2015) *BYOD –policy and practice in Higher Education in* Middleton, A., ed. (2015). Smart learning: Teaching and learning with smartphones and tablets in post compulsory education. Media-Enhanced Learning Special Interest Group and Sheffield Hallam University.

Vanwelsenaers, M. (2012). *Students using their own technology device in the classroom: Can "BYOD" increase motivation and learning.* Retrieved July, 30, 2014.

Veletsianos, G., & Navarrete, C. (2012). Online social networks as formal learning environments: Learner experiences and activities. *The International Review of Research in Open and Distributed Learning,* 13(1), 144-166.

Waterfield, J., West, B., & Parker, M. (2006). Supporting inclusive practice. In M. Adams, & S. Brown (Eds.), *Towards inclusive Learning in Higher Education: Developing Curricula for Disabled Students* (pp. 79-94). London: Routledge.

Welsh, K., & France, D. (2012). Spotlight on... Smartphones and fieldwork. *Geography*, *97*, 47.

Welsh, K. E., Mauchline, A. L., Park, J. R., Whalley, W. B., & France, D. (2013). Enhancing fieldwork learning with technology: practitioner's perspectives. *Journal of Geography in Higher Education*, *37*(3), 399-415.

Welsh, K.E, Mauchline, A.M, France, D., Park, J.R., Whalley, W.B., Powell, V. (submitted) Challenges of Bring Your Own Device (BYOD) for fieldwork. *Journal of Geography in Higher Education.*

Wood, A. (2012). BYOD: *The Pros and Cons for End Users and the Business.*

Woodcock, B. Middleton, A. Nortcliffe, A (2012) Considering the Smartphone Learner: an investigation into student interest in the use of personal technology to enhance their learning. *Student Engagement and Experience Journal Volume 1, Issue 1.*

Biographical Sketches

Servel Miller

Affiliation: Fellow Royal Geography Association;
Charter Geographer (GIS);
Fellow of the Higher Education Academy

Education: PhD., MSc., PgDip., BSc., FHEA, FRGS, CGeog
2008- 2010: University of Chester
PGCert in: Learning and Teaching in Higher Education
2003-2007: University of Liverpool- PhD in: GIS and Remote Sensing Application to Landslide Hazard management in North-East Wales.
2002-2003: Cranfield University
MSc in: Geographical Information Management (GIM)
International Institute for Aerospace Survey (ITC) Netherlands
1989-1994: University of the West Indies (UWI-Mona)
BSc in Geology

Business Address: University of Chester, Parkgate Road, Chester CH14BJ. UK

Research and Professional Experience:
2007- Present: Senior Lecturer & Programme Leader- University of Chester
2013-2016: Vice Chair National Flood Forum, UK
2004-2007: Visiting Lecturer - University of Chester

1994–2002: Senior Geologist - Mines and Geology Department, Jamiaca

Publications from the Last 3 Years:

Miller, S., Welsh, K. and France D, (2015) Development of Crisis and Emergency Risk Communication (CERC) Skills: Geoscience students Perception. *BELGEO* (1) *2015.*

Miller, S. Shalkowski, A, Harris, N., Richards, D. &Lyndon Brown, L. (2015) Landslides in Jamaica: Distribution, Cause, Impact and Management. *Encyclopedia of Natural Hazards*. Taylor Francis.

Miller, S. and Degg, M. (2015) Geographical Information Systems (GIS) applied to landslide hazard zonation in the North Wales coalfield. In Bevins, R., Nichol, D. & Bassett, M.G. (Eds), *Urban Geology in Wales: 4*. National Museum of Wales Geological Series No. 27, Cardiff.

Degg, M. and Miller, S. (2015) Landslide Hazard Mapping and Impact in the Holywell Area of NE Wales. In Bevins, R., Nichol, D. & Bassett, M.G. (Eds), *Urban Geology in Wales: 4*. National Museum of Wales Geological Series No. 27, Cardiff.

Morris, C., Miller, S. (2014) Coastal Landslide Mapping of the Black Ven Spittles Complex, Charmouth *Proceedings of the Dorset Natural History & Archaeological Society*, 01/2014; 135:165.

Miller, S. Leszczyńska, M., (2014) Evaluation of the use high resolution satellite imagery to map slope instability in a tropical environment: St. Thomas, Jamaica. *Vilnius Proceeding of 9th International Conference on Environmental Engineering*. Gediminas Technical University Press Technika.

Miller, S., Mandal, A. and Brown, L. (2014) Wave-emplaced boulders: implications for development of "prime real estate" seafront, North Coast Jamaica. *Bulletin of Engineering Geology and the Environment* 73(Issue1) Page p.109-122.

Katharine Welsh

Affiliation: University of Chester

Education: 2005-2009: PhD University of Liverpool
2002-2005: BSc University of Liverpool

Business Address: Parkgate Road, Chester, CH1 4BJ.

Research and Professional Experience:

2015 – present - Senior Lecturer at University of Chester
2010 – present – Visiting Lecturer at University of Chester
2013-2016 – Senior Researcher at University of Chester
2010-2013 – Postdoctoral Researcher at University of Chester
2009-2010 – Postdoctoral Researcher at University of Liverpool
2005-2009 – University Teaching Assistant at University of Liverpool

Professional Appointments: Fellow of Higher Education Academy

Publications from the Last 3 Years:

Schillereff, D. N., Chiverrell, R. C., Macdonald, N., Hooke, J. M., & Welsh, K. E. (2016). Quantifying system disturbance and recovery from historical mining-derived metal contamination at Brotherswater, northwest England. Journal of Paleolimnology, 56(2-3), 205-221.

Whalley, W. B., France, D., Mauchline, A., Welsh, K. & Park, J. (2016). Every student use of iPads: A vade mecum for students' active learning. In B. F. Baab, J. Bansavich, N. Souleles and F. Loizides (Eds.*), Proceedings of the 2nd International Conference on the Use of iPads in Higher Education*. Newcastle, United Kingdom: Cambridge Scholars Publishing.

France, D., Powell, V., Welsh, K.E., Mauchline, A.M, Park, J.R., Whalley, W.B. (2016) Ability of students to recognize the relationship between using mobile apps for learning during fieldwork and the development of graduate attributes. *Journal of Geography in Higher Education*, 40(2), p.182-192.

Miller, S., France, D., Welsh, K.E. (2016) *Development of Crisis and Emergency Risk Communication (CERC) Skills: Geoscience students Perception*. Belgeo.

Welsh, K.E., Mauchline, A., Powell, V., France, D., Park, J.R., Whalley, W.B. (2015) Student perceptions of iPads as mobile learning devices for fieldwork. *Journal of Geography in Higher Education*, 39(3), p.450-469.

France, D., Whalley, W.B., Welsh, K.E., Powell, V., Lerczak, A., Mauchline, A., Park, J.R. (2015*) Enhancing Fieldwork Learning using Mobile Technologies*. Springer Brief Series.

Whalley, B. W., France, D., Park, J. R., Mauchline, A. L., Powell, V. and Welsh, K. E.(2014) iPad use in fieldwork: formal and informal use to enhance pedagogical practice in a Bring Your Own Technology world. In: *First International Conference on the use of iPads in Higher Education*, 20-22 March 2014, Paphos, Cyprus, pp. 100-120.

In: Mobile Learning
Editors: D. René and C. Aubin
ISBN: 978-1-53612-801-7

Chapter 8

THE OPPORTUNITIES AND CHALLENGES OF MOBILE GAME-BASED LEARNING

Judith Ramsay[1,*], Melody M. Terras[2] and Elizabeth A. Boyle[1]
[1]Department of Interdisciplinary Studies,
Manchester Metropolitan University, Crewe, UK
[2]School of Media, Culture and Society,
University of the West of Scotland,
Paisley, UK

ABSTRACT

Mobile learning has witnessed increasing adoption across a broad spectrum of educational domains. In recent years, however, a number of psychological challenges have been identified that influence the relative effectiveness of mobile learning. These include the limited nature of human information processing, including working memory capacity and the ability to attend to stimuli, especially in mobile environments that constantly change (Terras and Ramsay, 2012). The use of games-based learning has also increased. Whilst it is notable that in recent years the number of mobile games in particular has increased, less is known about

* Corresponding Author E-mail: j.ramsay@mmu.ac.uk.

the psychological challenges they specifically present. Recent work indicates the importance of considering the psycho-social profile of the mobile game player, the context and mechanics of the game (Boyle, Terras and Ramsay, 2017). In this expert commentary, we build upon previous research into the psycho-social profile of the learner to develop a rich, nuanced understanding of the specific socio-emotional factors that influence mobile game play: these include player emotion, motivation, meta-cognition and the importance of context and how to learn, using games, across changing temporal and spatial contexts. We then integrate this psychological understanding with the literature on game mechanics along with the interaction between the mechanics of the game and mobile learner characteristics. We argue that it is through developing an insight into the intersection of mobile learning and mobile gameplay that significant gains will be made in the quest to develop effective mobile games for learning.

1. Introduction

Rapid technological advances are substantially influencing the learning landscape, offering learners, providers and developers a range of new platforms for learning such as Web 2.0 and the early signs of Web 3.0, mobile learning and game-based learning. Whilst game-based learning is expanding (Boyle et al., 2016) game-based mobile learning (GBML) has received less research attention and will serve as the focus of this expert commentary. Mobile gaming is a highly popular leisure activity and Holleis, Sahami and Schmidt (2011) outline three distinctions between mobile and traditional games, namely that they are used to both fill and to kill time (they cite the example of waiting for public transport), and are played across different contexts. However, there is increasing interest too in using mobile games for learning. Whilst digital game-based learning comprises an educational component and an entertainment component (Bellotti et al., 2013), in this chapter we characterise and explore the distinct nature of digital game-based learning *whilst mobile.* It the use across different contexts that is key, as noted by Holleis et al., (2011). Boyle et al., (2017) discussed the importance of considering the psycho-social profile of the game player, the context, and the mechanics of the game in order to maximise the formal and informal learning opportunities of both entertainment and serious games. Therefore, this expert commentary will consider the influence of these factors on the success of mobile game-based learning. We propose that the educational benefits of game-based mobile learning can best be maximised by taking an integrated perspective that draws together insights gained from the currently relatively distinct field of mobile learning, in particular the insights that can be gained from a psychological perspective (Terras and Ramsay, 2017). These should be considered together with game-focused research to inform a holistic understanding of the challenges and opportunities presented by mobile game-based learning.

1.1. The Mobile Learning Opportunity

Mobile learning has witnessed increasing uptake across a broad spectrum of educational domains such as history (Akkerman, Admiraal and Huizenga, 2009), science (Zydney and Warner, 2016) and language learning (Huang et al., 2016). Definitions of mobile learning and the opportunity afforded have developed in step with technological advances. Its definition has evolved from characterising this form of learning in terms of technological enablement, to encompass the central human dimensions that frame mobile learning. Traxler (2007) declared that "mobile" had come to refer less to the way in which one learns i.e., on the move, rather, that it had far deeper consequences for learning per se. It has since been defined as "understanding and knowing how to utilize our everyday life-worlds as learning spaces" (Pachler, Bachmair and Cook, 2010, p. 6). More recently, Kukulska-Hulme (2013) emphasised the importance of mobile learner characteristics and identified different types of mobile learner. Here, we define mobile learning as "the use of wireless handheld devices to support learning and teaching across different psychological, social, physical and temporal contexts" (Terras and Ramsay, 2017, p. 6398). It is by adopting this wide definition, which recognises the need to leverage an understanding of human behavior, that the mobile gaming learning opportunity can be maximised.

1.2. The Learning Opportunities of Mobile Games

Game-based mobile learning (GBML) refers to a specific form of mobile learning that utilises digital games as the learning vehicle. Mobile games include engaging features that are typical of digital games generally, e.g., rewards, levels, progress, pacing, rules, competition and feedback. However, it is also useful to consider characteristics that may be unique to, more relevant to or distinct to mobile games such as location sensitivity, feedback, personalization and use of multimedia. Furthermore, mobile games bring further benefits as they are uniquely placed to support anytime and anywhere access that offers great potential for contextualized learning and decision-making in real world contexts.

Both entertainment games and serious games have capitalised upon the unique location features available in mobile games such as the ability to use or progress through different locations: Pokémon Go uses satellite navigation (SATNAV) technology to present virtual creatures (Pokémon) that users interact with as if they were in the same real-world location as the player. Historical or cultural games involving storification can promote the learning of history using mobile games (Akkerman et al., 2009). A range of nature and natural science-based games and applications (Huang, Lin and Cheng, 2010) draw upon Global Positioning System (GPS) to promote location awareness and allow the location of both the user and learning objects and resources to be tracked, guided and the route monitored. This supports increased outdoor learning and increased academic achievement (Chiang, Yang and Hwang, 2014). When supported by location-based awareness, mobile games can facilitate real world, in-the-moment learning which is otherwise very challenging to provide. As well as location information, mobile games can also use contextual information to provide game activities. For example, Pollak et al., (2010) described a game that encouraged children to become more aware of healthy eating. Players posted photos of the food that they were eating that day and their photos were scored with respect to how healthy the food was. The aim of the game was to eat healthy food in order to keep a virtual pet alive and the study provides a good illustration of how mobile games provide an engaging platform for health-related initiatives. Another prevalent feature of mobile games is the feedback provided to players. Zydney and Warner (2016) provide a comprehensive summary of how feedback in mobile games for science provides scaffolding for learning, i.e., timely support, different ways - conceptual, metacognitive, procedural and strategic. Conceptual scaffolds provide support for clarifying understanding of concepts as demonstrated by the work of Chu et al., (2010) who provided support for players in categorising plants. Metacognitive scaffolding encourages players to self-regulate, i.e., to reflect on what and how they are learning, and can offer guidance to help them to monitor and manage their learning (Looi et al., 2014). Strategic scaffolding provides advice about the most appropriate strategy to use at a specific point in a game. For example, in Ahmed and Parson's (2013) ThinknLearn game, players were given prompts about specific actions that were required at different points in a cycle of activities, such as being asked to take measurements in the exploration phase, or select a possible hypothesis in the explanation stage. Further examples of how mobile games offer excellent opportunities for feedback is demonstrated in mobile language games such as Duolingo (https://www.duolingo.com/mobile) and Babbel (https://uk.babbel.com/mobile) which are very effective as they can provide prompt feedback about the appropriateness of players' vocabulary and grammar. Personalisation is an important feature of mobile games as they increasingly have the capacity to collect data from players, and this can then be used to provide them with personalised feedback. Data collected includes physiological data, duration and accuracy of players' responses, but also personal information and behavioural data. Many people have experience of this personalisation of data in the form of fitness trackers and applications that record and collate the number of steps taken, calories consumed, heart rate and sleep patterns. This is then fed back to users to support them in adapting their health behaviours. Ventola (2014) describes how the use of mobile devices and applications are transforming health care practice by supporting clinical decision-making and increasing efficiency. Mobile games make good use of multimedia stimuli that assist the learning process, in terms of presenting information and supporting content and knowledge representations, via concept maps, annotated visual representations and animations (Zydney and Warner, 2016). These contextualised aspects have resulted in the increase of mobile game-based learning across a range of formal and informal learning contexts (Koutromanos and Avraamidou, 2014), especially in relation to science learning (Li and Tsai, 2013). Despite their increasing uptake, a recent review of the educational effectiveness of mobile games identified areas for future research e.g., the need to make more timely use of technological advances, a finer-grained level of evaluation of specific design features and increased integration between theory and design (Zydney and Warner, 2016). Many of these recommendations are addressed in the following sections of this expert commentary as we discuss how the benefits of game-based learning can be maximised by adopting a more psychologically-informed approach. They highlight not only the importance of considering learner characteristics but also how they interact with game mechanics and the context of use.

2. Viewing Mobile Game-Based Learning through a Psychological Lens

The fact that mobile devices are often heavily personalised by the user, and support a blend of both formal and informal activities, increases the likelihood of learners using such devices to support their learning in an individually-tailored opportunistic way. Such personalisation highlights the need for a full understanding of user behavior and an appreciation of the fact that the increasing use of mobile devices for learning brings not only opportunities but a range of practical and psychological challenges (Terras and Ramsay, 2012; 2014a). These challenges, which encompass every form of mobile learning, need to be addressed by educators, instructional designers and software developers alike. The generic challenges for mobile learning are as follows. Firstly, the limited nature of human information processing, namely our finite working memory capacity and (in)ability to attend to stimuli, especially in mobile environments that constantly change may negatively impact upon learning. This is compounded by the increased susceptibility to interruptions as a direct consequence of learning in contexts that change and that are not explicitly designed as "learning environments" in the traditional sense. Additionally, human memory is context-dependent. As contexts change during mobility, there is a consequent, accompanying risk to reliable memory encoding and retrieval. Learning is distributed and it is situated (Suchman, 1987; 2007); indeed, location-based games recognise this and exploit GPS data that can be integrated into a learner's game to enable situated learning (Scepanovic et al., 2016). Finally, need to develop the psychological infrastructure for mobile learning, namely, the challenge is for mobile learners to become aware of and manage their cognitive resources (metacognition) when learning whilst mobile.

2.1. The Psychological Challenges of Mobile Game-Based Learning

As mentioned previously, these psychological challenges hold true for all forms of mobile learning, including mobile game-based learning. However, we propose that mobile game-based learning comes with its own specific set of challenges that exist in addition to the aforementioned generic ones. Whilst game-based learning is sometimes considered to have a unique nature, requiring specialist skills (of both developers and gamers), the rapid adoption of this form of learning strongly indicates that research into learner preferences and skills lies at the heart of the quest to support this form of learning. We propose that this can best be achieved by adopting a psychologically-informed perspective in the first instance.

2.2. Changing Spatio-Temporal Contexts

Contextual change presents a major challenge to mobile learning. We propose that it essential to recognize the multi-faceted nature of the mobile game-based learning context. Firstly, it is necessary to distinguish between location and context. From a traditional mobile learning perspective, context is typically regarded as being external to the learner and predominantly reflects the location/environment the learner inhabits at any one point in time. From a psychologically-informed perspective, context is conceptualised as being both external to the individual in the traditional sense but also internal in that it reflects the cognitive and emotional state of the learner (Terras and Ramsay, 2014b). When a psychological perspective is applied to games, context has three facets: (1) the physical external context; (2) the internal (psychological) context and (3) the context created by the game itself (Boyle et al., 2017).

It is also essential to recognise the temporal dimension of the location and the external, internal and games-based context as each varies and evolves over time e.g., people may enter or leave the physical location bringing with them their own type of distraction, and players may enter and leave the game and the difficulty of the game may vary over time. However, time is not simply physical, it is also psychological (Zakay, 2012). It is particular to each individual, is context-dependent, non-linear, and is shaped by social, cognitive and emotional factors. For example, the Attentional Gate Model (Zakay, 2000) indicates that the allocation of attentional resources is determined by aspects of the temporal context that are very specific e.g., waiting for a signal or an alert. The importance of the psychological experience of time is receiving increased recognition in the wider mobile learning literature (Terras and Ramsay, 2014b) and we propose that this is essential to the understanding of contextual influences on mobile game-based learning. The experiential perception of time may offer insights into the experience of immersion and flow within games. Such insights may also assist understanding of individual differences in experience and also inform the development of the mechanics of the game itself. A consideration of the individual differences captured by the learner's internal psychological context is presented in the following section.

2.3. The Psycho-Social Profile of the Learner/Gamer

The current evidence base concerning individual differences with respect to mobile game-based learning is rather limited, so here we draw upon the existing evidence base from games generally. Existing research highlights the importance of motivation and emotion. Playing games has a beneficial effect upon one's emotional state (Ryan et al., 2006), with game players experiencing more positive than negative emotions when encountering failure (Salminen and Ravaja, 2008). It has also been demonstrated that engagement in game playing promotes learning and that the experience of challenge in turn promotes engagement (Hamari et al., 2016). Mobile game-based learning brings the additional motivational advantages of games (Chen, Liu and Hwang, 2016) and Boyle et al., (2012) argue that three theories may explain why individuals play digital games. Firstly, self-determination theory (Deci and Ryan, 1985) which proposes that game playing is driven by the need to feel autonomy, competence and relatedness. For example, Ryan, Rigby and Przybylski (2006) found that self-determination theory was instrumental in predicting player's enjoyment of games along with player presence. Secondly, the Technology Acceptance Model (TAM) (Davis, 1989) proposes that ease of use and usefulness determine technology use. Uses and gratifications theory (Katz, Blumler and Gurevitch, 1973; Ruggerio, 2000) by comparison holds that a variety of media are adopted to meet entertainment needs. For example, Chou and Tsai (2007) demonstrated that games served entertainment needs, information-seeking needs and social needs. Chen et al., (2016) demonstrated that it is possible to increase students' motivation to learn when using games through an effective gaming mechanism and multi-stage guiding mechanism, thus drawing on an understanding that motivation is highly instrumental in using games for learning.

Playing games has also been demonstrated to assist in the re-evaluation of one's ability to manage or cope with a situation (Gross and John, 2003). Gamers who engage with games on a more regular basis have been found to regulate their own emotional states more than less regular players (Gaetan, Bréjard and Bonnet, 2016). We agree with Kukulska-Hulme and Jones (2011) who have argued that an integrative approach needs to be adopted across areas of research endeavour, incorporating the context of learning, technologies and individual learner practices. In particular, Kukulska-Hulme (2013) has more recently highlighted the importance of recognising that there are very different types of mobile learner, which complements the perspective taken by Terras and Ramsay (2012; 2014b) that individual differences play a key part in mobile learning.

One of the key individual differences to emerge from the literature lies at the intersection of 'flow' - the experience of being absorbed in an activity to the extent that self-awareness disappears - and motivation. The subjective individual experience of flow (Csikszentmihalyi and Csikszentmihalyi, 1990; Csikszentmihalyi, 1993) has been demonstrably influential in engagement with digital games, with the consensus being that flow is a pleasurable state akin to enjoyment. Enjoyment, in turn has been demonstrated to be an emotion that plays an important role in computer game playing (Fang and Zhao, 2010), with intrinsic motivation in particular emerging as a key predictor in the enjoyment of games (Baek and Touati, 2017). More broadly, a sophisticated understanding of learner characteristics that includes their media literacy skills is required, along with their ability to engage effectively with others in online learning contexts (Ramsay and Terras, 2015). Research endeavor now needs to ascertain the extent to which these factors transfer across to game-based mobile learning in its myriad forms.

2.4. Meeting the Challenges: The Mobile Game-based Learner in Context

Arguably, the most important factor that is instrumental in meeting the challenges set out above is in ascertaining the specific dimensions of context that influence mobile game-based learning, in particular those that shape the interaction between the context (in all of its different forms) and the learner. One important aspect of context, for example, is the extent to which the learner's screen time is at their own discretion or controlled by others across their contexts of use (e.g., by parents and siblings) as this shapes behaviors (Terras and Ramsay, 2016). Despite the challenges associated with changing spatio-temporal contexts, a major advantage of mobile learning is its ability to support anytime and especially anywhere access that offers great potential for contextualized learning and decision-making in real world contexts. As mentioned earlier, with regards to game mechanics, both entertainment games and serious games exploit location features, however, the potential interactions are even more complicated when it comes to mobile game-based learning as context plays a more important role as the location and context of the user is often a key aspect of the learning process. Moreover, future research must identify not only those contextual aspects that are most influential but how and when the respective aspects interact. For example, is the game player's 'real world' locational context (and the social and physical elements within in) as influential or as the 'designed' context of the game itself, or vice versa? The relevant importance of each aspect of context may also vary depending on the nature and purpose of the games, with highly situated games drawing more heavily upon external contexts. What is ultimately required, beyond a consideration of the interactions between user characteristics and context is a sophisticated characterisation of how the aforementioned, is, in turn, influenced by game mechanics. Over the past decades, specifications and standards for the design of mobile devices and applications have matured and been subject to widespread implementation. Nevertheless, interpretations exist in the way in which game mechanics are *captured* (what is it that the game needs to do), *specified* (how the functionality is expressed on paper) and *rendered* (how it is designed across both software and hardware components). All of these serve in combination to influence the manner in which the game is played. By extension, it also influences the extent of the scaffolding that is provided to the game player which is turn may help to determine the effectiveness of the learning achieved as well as the motivation to continue and enjoy the game. Additionally, game mechanics play an important role in the learner's experience of flow. Bressler and Bodzin (2013) examined flow in mobile augmented reality games and found that players' attitude to gaming was the best predictor of flow experience, as compared with gender or interest in science, neither of which predicted flow experience.

Conclusion and Recommendations for the Future

In conclusion, the importance of considering psychological differences is becoming more apparent in recent research e.g., in the exploration of time perception, multitasking and flow (Chinchanachokchai, Duff and Sar, 2015). This is a good example of research that integrates different perspectives, time perception and multitasking. This emergent area of research interest is focused on understanding the influence of user characteristics in context and how these factors influence behavior and the ability to process information. This aim may best be realised with increased multi-disciplinary working and the adoption of a more integrated approach as advocated by Boyle, Terras and Ramsay (2017) who identified context, the psychosocial profile of the game player and game mechanics as key dimensions for integrating game-based learning research within the wider context of e-learning.

If future multi-disciplinary research is to be successful then it needs to be more theoretically- driven. Much of the existing research on mobile game-based learning is somewhat atheoretical in nature or draws on a range of different theoretical perspectives including situated learning theory, inquiry-based learning and socio-cultural theory (Zydney and Warner, 2016). It is interesting to note that many, if not all of the theoretical perspectives adopted tend to be contextually-situated. This is not unexpected given the importance of context in mobile game-based learning. However, future research would benefit from increased and more explicit recognition of the contextualised nature of learning. As discussed earlier, context is complex and multi-faceted and changes over time, so the design of mobile learning games needs to be highly sensitive to context and ensure that appropriate learning support is included in the game to maximise the learning opportunities and minimise the distractions that context brings.

If the benefits of game-based learning are to be fully realised that then they need to draw upon technological advances in a time-efficient manner. Current applications tend to be slightly behind the curve (Zydney and Warner, 2016). Keeping apace with technological developments is challenging and can benefit from examining research concerning the nature of the psychological profile of the learner - which does not evolve as quickly as technology. There is currently much excitement about augmented reality and immersive technology and human information overload has been identified as a major challenge (Wu et al., 2013). Interestingly, they also advocate the need to examine instructional approaches and highlight the importance of "alignment among technology design, instructional approach, and learning experiences" (p. 41) and identify "three categories of instructional approaches that emphasize the "roles," "tasks," and "locations," (p. 41) and "discuss what and how different categories of augmented reality approaches may help students learn."

A further aspect of technological development is the increasing availability and use of different devices. Learning whilst moving through different contexts and times increases the likelihood that the learner will use different digital media in different locations and times to access games to continue their learning. Such usage has been termed "trans-media" (e.g., Lindt et al., 2007) raising the question of whether the nature of the access device has an impact on learning. Pervasive games (Valente et al, 2017), can be defined as "context aware games that use mobile devices" (Valente et al., 2017, p. 137). Whilst the quality-related requirements of pervasive games have been researched and to a great extent captured (Valente et al., 2017), the human behavioral requirements remain to be determined. As stated earlier, of particular relevance is the internal context, the internal psychological state of the mobile games based learner and the attendant opportunities and challenges it presents. The human psychological limitations that have long been established in the field of Psychology have remained more or less constant since the advent of large screen PCs, however the increasingly small devices humans are currently using are placing ever greater pressures on our limited psychological (and motor) resources. Psychological models of human visual processing and attention, Fitts' Law (1954) regarding the time taken to access a target object, and that tracking of patterns of fixation tell us much about human capabilities and their attendant limitations. These limitations are now increasingly faced with the fresh suite of constraints that come with mobile device screen use such as small visual displays, short session lengths and greater reliance upon working memory capacity to remember information that is "off screen."

In summary:

- Taking a psychological perspective can yield important insights into the opportunities and challenges associated with mobile game-based learning.
- Within this, the adoption of an integrated approach that considers the learner, the learning context and the interaction between the learner and the mechanics of the game is particularly illuminating.
- Whilst we have highlighted the insights that can be gained from a psychological perspective, future theory, research and practice should be not only multi-disciplinary but interdisciplinary if it is to truly yield up the mechanisms that facilitate effective mobile game-based learning.

REFERENCES

Ahmed, S., & Parsons, D. (2013). Abductive science inquiry using mobile devices in the classroom. *Computers and Education, 63,* 62-72.

Akkerman, S., Admiraal, W., & Huizenga, J. (2009). Storification in history education: a mobile game in and about medieval Amsterdam. *Computers and Education, 52*(2), 449-459.

Babbel Accessed 6th August 2017 from https://uk.babbel.com/mobile.

Baek, Y., & Touati, A. (2017). Exploring how individual traits influence enjoyment in a mobile learning game. *Computers in Human Behavior, 69,* 347-357.

Bellotti, F., Kapralos, B., Lee, K., Moreno-Ger, P., & Berta, R. (2013). Assessment in and of serious games: an overview. *Advances in Human-Computer Interaction, 2013,* 1.

Boyle, E. A., Connolly, T. M., Hainey, T., &Boyle, J. M. (2012). Engagement in digital entertainment games: a systematic review. *Computers in Human Behavior, 28,* 771–780.

Boyle, E. A., Hainey, T., Connolly, T. M., Gray, G., Earp, J., Ott, M., Lim, T., Ninaus, M., Ribeiro, C., & Pereira, J. (2016). An update to the systematic literature review of empirical evidence of the impacts and outcomes of computer games and serious games. *Computers and Education, 94,* 178-192.

Boyle, E. A., Terras, M. M., & Ramsay, J. (2017). The Challenges and Opportunities of Entertainment Games and Serious Games for formal and informal learning. In S. Rutherford (Ed.) *Informal Learning: Perspectives, Challenges and Opportunities.* New York: Nova Science Publishers.

Bressler, D. M., & Bodzin, A. M. (2013). A mixed methods assessment of students' flow experiences during a mobile augmented reality science game. *Journal of Computer Assisted Learning, 29*(6), 505-517.

Chen, C. H., Liu, G. Z., & Hwang, G. J. (2016). Interaction between gaming and multistage guiding strategies on students' field trip mobile learning performance and motivation. *British Journal of Educational Technology, 47*(6), 1032-1050.

Chiang, T. H., Yang, S. J., & Hwang, G. J. (2014). An augmented reality-based mobile learning system to improve students' learning achievements and motivations in natural science inquiry activities. *Journal of Educational Technology and Society, 17*(4), 352.

Chou, C., & Tsai, M.-J. (2007). Gender differences in Taiwan high school students' computer game playing. *Computers in Human Behavior, 23,* 812–824.

Chu, H. C., Hwang, G. J., Tsai, C. C., & Tseng, J. C. R. (2010). A two-tier test approach to developing location-aware mobile learning systems for natural science courses. *Computers and Education, 55*(4), 1618-1627.

Chinchanachokchai, S., Duff, B. R., & Sar, S. (2015). The effect of multitasking on time perception, enjoyment, and ad evaluation. *Computers in Human Behavior*, *45*, 185-191.

Csikszentmihalyi, M. (1993). *The evolving self: A psychology for the third millennium.* New York, NY: Harper Collins.

Csikszentmihalyi, M., & Csikszentmihalyi, I. (1990). *Flow: The psychology of optimal experience.* New York, NY: Harper and Row.

Davis, F. D. (1989). Perceived usefulness, perceived ease of use, and user acceptance of information technology. *MIS Quarterly, 13,* 319–339.

Deci, E. L., & Ryan, R. M. (1985). *Intrinsic motivation and self-determination in human behaviour*. New York: Plenum.

Duolingo Accessed 6th August 2017 from https://www.duolingo.com/mobile.

Fang, Z., & Zhao, F. (2010). Personality and enjoyment of computer game play. *Computers in Industry, 61*, 342–349.

Fitts, P. M. (1954). The information capacity of the human motor system in controlling the amplitude of movement. *Journal of Experimental Psychology, 47,* 381-391.

Gaetan, S, Bréjard, V., & Bonnet, A. (2016). Video games in adolescence and emotional functioning: Emotion regulation, emotion intensity, emotion expression, and alexithymia. *Computers in Human Behavior, 61,* 344-349.

Gross, J. J., & John, O. P. (2003). Individual differences in two emotion regulation processes: Implications for affect, relationships, and wellbeing. *Journal of Personality and Social Psychology, 85,* 348-362.

Hamari, J., Shernoff, D. J., Rowe, E., Coller. B., & Edwards, T. (2016). Challenging games help students learn: An empirical study on engagement, flow and immersion in game-based learning. *Computers in Human Behavior*, 54, 170-179.

Huang, Y. M., Lin, Y. T., & Cheng, S. C. (2010). Effectiveness of a mobile plant learning system in a science curriculum in Taiwanese elementary education. *Computers and Education, 54*(1), 47-58.

Holleis, P, Sahami A., & Schmidt, A. (2011). Providing Generic Context for Mobile Games on Phones. In M. Beigl, H. Christiansen, T. R. Roth-Berghofer, A. Kofod-Petersen, K. R. Coventry, & H. R. Schmidtke (Eds), *Modeling and Using Context. CONTEXT 2011. Lecture Notes in Computer Science, 6967.* Springer, Berlin, Heidelberg.

Katz, E., Blumler, J. G., & Gurevitch, M. (1973). Uses and Gratifications Research. *Public Opinion Quarterly*, *37,* 509-523.

Koutromanos, G., & Avraamidou, L. (2014). The use of mobile games in formal and informal learning environments: a review of the literature. *Educational Media International*, *51*(1), 49-65.

Kukulska-Hulme, A. (2013). Mobile learners: who are they and who will they become? In Z. Berge and L. Y. Muilenburg (Eds.), *Handbook of Mobile Learning* (pp. 145–154.). New York: Routledge.

Kukulska-Hulme, A., & Jones, C. (2011). The next generation: design and the infrastructure for learning in a mobile and networked world. In A. D. Olofsson and J. Lindberg (Eds.), *Informed Design of Educational Technologies in Higher Education: Enhanced Learning and Teaching* (pp. 57–78). Hershey, PA: Information Science Reference (an Imprint of IGI Global).

Li, M. C., & Tsai, C. C. (2013). Game-based learning in science education: A review of relevant research. *Journal of Science Education and Technology*, *22*(6), 877-898.

Lindt, I., Ohlenburg, J., Pankoke-Babatz, U., & Ghellal, S. (2007). A report on the crossmedia game epidemic menace. *ComputEntertain CIE* 5(1):8.

Looi, C. K., Sun, D., Wu, L., Seow, P., Chia, G., Wong, L. H., Soloway, E., & Norris, C. (2014). Implementing mobile learning curricula in a grade level: empirical study of learning effectiveness at scale. *Computers and Education, 77,* 101-115.

Pachler, N., Bachmair, B., & Cook, J. (2010). *Mobile learning: structures, agency, practices.* New York: Springer.

Pollak, J., Gay, G., Byrne, S., Wagner, E., Retelny, D., & Humphreys, L. (2010). It's time to eat! Using mobile games to promote healthy eating. *IEEE Pervasive Computing, 9*(3), 21-27.

Ramsay, J., & Terras, M. M. (2015). The Pendulum Swing of User Instruction and Interaction: Resurrection of "How to Use" Technology to Learn in the 21st Century. *E-Learning and Digital Media. May-July 2015,* (12), 372-390.

Ruggiero, T. E. (2000). Uses and Gratifications Theory in the 21st Century. *Mass Communication and Society, 3*(1), 3-37.

Ryan, R. M., Rigby, C. S., & Przybylski, A. (2006). The motivational pull of video games: A self-determination theory approach. *Motivation and Emotion, 30,* 347–363.

Salminen, M., & Ravaja, N. (2008). Increased oscillatory theta activation evoked by violent digital game events. *Neuroscience Letters, 435*, 69–72.

Scepanovic, S, Vujicic, T, Radunović, P, Antovic, I., & Vlajic, S. (2016). Challenges in design of games for mobile learning. *10th International Technology, Education and Development Conference 7-9 March, Valencia, Spain* pp. 2989-2998.

Suchman, L. A. (1987). *Plans and Situated Actions: The Problem of Human-Machine Communication*. Cambridge: Cambridge Press.

Suchman, L. A. (2007). *Human-machine reconfigurations.* New York: Cambridge University Press.

Terras, M., & Ramsay, J. (2012). The five central psychological challenges facing effective mobile learning. *British Journal of Educational Technology, 43*(5), 707–832.

Terras, M. M., & Ramsay, J. (2014a). A Psychological Perspective on the Temporal Dimensions of E-learning. *E-Learning and Digital Media, 11*(2), 108-132.

Terras, M. M., & Ramsay, J. (2014b). E-learning, Mobility and Time: A Psychological Framework. In E. Barbera and P. Reimann (Eds.),

Assessment and Evaluation of Time Factors in Online Teaching and Learning. Hershey: IGI Global Publishing.

Terras, M. M., & Ramsay, J. (2016). Family Digital Literacy Practices and Children's Mobile Phone Use. *Frontiers in Psychology: Personality and Social Psychology*, 7, 1957.

Terras, M. M., & Ramsay, J. (2017). A Psychological Perspective on Mobile Learning. In M. Khosrow-Pour (Ed.), *Encyclopedia of Information Science and Technology*, 4th edition. Hershey: IGI Global Publishing.

Traxler, J. (2007). Defining, Discussing, and Evaluating Mobile Learning: The moving finger writes and having writ... (International Review of Research in Open and Distance Learning, 8(2), June 2007). Accessed on 16/06/2017 from http://www.irrodl.org/index.php/irrodl/article/viewArticle/346/875oor

Valente, L., Feijo′, B., Sampaio, J. C., & do Prado Leite, J. C. S. (2017). Mapping quality requirements for pervasive mobile games. *Requirements Engineering, 22,* 137–165.

Ventola, C. L. (2014). Mobile devices and apps for health care professionals: uses and benefits. *Pharmacy and Therapeutics, 39*(5), 356-364.

Wu, H. K., Lee, S. W. Y., Chang, H. Y., & Liang, J. C. (2013). Current status, opportunities and challenges of augmented reality in education. *Computers and Education, 62*, 41-49.

Zakay D. (2000). Gating or switching? Gating is a better model of prospective timing. *Behavioural Processes, 50,* 1–7.

Zakay, D. (2012). Experiencing time in daily life. *The Psychologist, 25*(8), 578–581.

Zydney, J. M., & Warner, Z. (2016). Mobile apps for science learning: Review of research. *Computers and Education, 94*, 1-17.

INDEX

A

acceptance, v, vii, ix, 4, 61, 62, 63, 65, 66, 68, 74, 78, 81, 83, 93, 97, 98, 114, 116, 118, 119, 180
access, viii, x, 11, 12, 22, 23, 33, 38, 42, 49, 62, 64, 69, 72, 74, 79, 83, 85, 86, 89, 90, 94, 96, 123, 126, 127, 136, 138, 141, 143, 149, 150, 153, 154, 168, 175, 177
access device, 177
achievement test, 108, 109, 110
active feedback, 136
active learning, 42, 54, 164
age, 26, 31, 37, 42, 63, 66, 67, 68, 74, 78, 134
alexithymia, 180
app, xi, 2, 42, 43, 44, 45, 74, 77, 79, 83, 86, 88, 89, 90, 91, 92, 93, 94, 95, 96, 98, 102, 103, 104, 109, 154, 157, 162, 172
artificial intelligence, 117, 119
assessment, x, 6, 42, 43, 56, 66, 80, 106, 107, 116, 179
assessment tools, 43
assistive technology, 157
attitudes, 5, 22, 135, 143, 147, 158

B

benefits, ix, 4, 34, 35, 42, 135, 136, 137, 140, 143, 147, 167, 168, 170, 177, 183
blended learning, v, x, 79, 80, 81, 83, 88, 91, 92, 96, 97, 116, 118, 131
blogs, 97, 98
brainstorming, 85
bring your own device (BYOD), vi, vii, viii, xii, 34, 38, 127, 133, 134, 135, 136, 137, 138, 139, 140, 141, 143, 145, 146, 147, 148, 149, 150, 151, 152, 153, 154, 155, 156, 157, 158, 159, 160
browser, 86

C

challenges, vii, xii, xiii, 5, 7, 19, 38, 45, 46, 47, 52, 53, 131, 133, 135, 153, 159, 165, 167, 170, 171, 174, 177, 178, 182, 183
children, 3, 4, 5, 83, 168
classes, xii, 22, 27, 28, 29, 36, 38, 41, 43, 44, 46, 47, 48, 53, 122, 152, 157
classic literature, v, vii, 1, 2, 4
classroom, vii, viii, xi, xii, 8, 10, 28, 30, 33, 34, 39, 41, 42, 44, 46, 47, 51, 54, 101,

102, 103, 104, 105, 106, 108, 109, 110, 112, 113, 114, 115, 116, 124, 125, 126, 127, 128, 129, 133, 135, 136, 137, 138, 139, 140, 144, 146, 147, 148, 153, 154, 157, 159, 160,178
cloud computing, 130
communication, 38, 40, 62, 65, 73, 123, 129, 136
communication technologies, 40, 62
compulsory education, 160
computer, 3, 4, 5, 8, 23, 31, 43, 47, 62, 105, 113, 114, 117, 119, 123, 124, 126, 150, 174, 179, 180
computer games, 3, 4, 5, 8, 31, 179
computer technology, 114
computing, 40, 63, 64, 74, 83, 134, 156, 160
cost, xii, 82, 133, 134, 137, 140, 154
counter measures, 92
creativity, 37, 130, 150
critical analysis, xi, 121
critical thinking, 115, 118
culture, 2, 3, 31, 40, 82
curricula, 143, 181
curriculum, 7, 8, 39, 42, 48, 102, 114, 125, 181
cyberculture, 39, 40, 54, 55

D

data collection, ix, 34
data mining, 119
diagnosis, vi, xi, 101, 102, 103, 105, 106, 107, 112, 113, 117, 118, 119
digital games, 2, 3, 4, 5, 6, 7, 8, 23, 39, 168, 173, 174
digital technologies, 43, 47, 57, 122, 124, 127, 134
Diminuto, 43, 44

E

education, vii, viii, ix, xi, xii, 1, 2, 3, 4, 6, 7, 8, 10, 31, 33, 34, 36, 37, 39, 40, 45, 53, 56, 62, 64, 66, 67, 68, 74, 75, 77, 80, 102, 104, 108, 112, 114, 121, 122, 126, 127, 130, 131, 134, 135, 155, 156, 157, 178, 181, 183
educational assessment, 114
educational design, 122
educational objective, 7, 9, 37
educational opportunities, 36
educational practices, 34, 36, 39
educational process, 125
educational scenario, 38
educational system, 128
educators, 3, 130, 139, 170
e-learning, vii, x, 29, 31, 54, 76, 79, 80, 83, 86, 87, 88, 94, 96, 116, 122, 126, 159, 176, 182
elementary school, 104, 106
emotion, xiii, 166, 173, 174, 180
environment, xii, 6, 8, 9, 46, 64, 77, 93, 97, 103, 105, 113, 127, 131, 133, 134, 162, 172
evaluation process, 43
everyday life, xi, 83, 121, 124, 167
extrinsic motivation, 6, 30

F

flipped classroom, vi, vii, xi, 101, 102, 103, 104, 105, 106, 108, 109, 110, 112, 113, 114, 115, 116
foreign language, vii, ix, 28, 34, 37, 52, 53, 56
formal education, 37, 73
formative process, 122, 123

G

game-based learning, v, vi, vii, viii, ix, 1, 2, 4, 5, 6, 10, 30, 31, 34, 165, 166, 170, 171, 172, 173, 174, 176, 177, 178, 180
games, xiii, 2, 3, 4, 5, 6, 7, 8, 9, 10, 19, 22, 23, 25, 26, 28, 29, 30, 31, 45, 46, 47, 49, 54, 69, 71, 72, 73, 165, 166, 168, 171, 172, 173, 174, 175, 176, 177, 179, 180, 181, 182, 183
games in education, 3, 7
gamification, ix, 10, 34, 45, 46, 47, 53, 54, 55, 56
gaming-based learning platforms, 42
Google, 22, 38, 42, 43, 48

H

handheld devices, 63, 64, 167
higher education, v, vi, ix, xii, 55, 58, 59, 61, 62, 65, 66, 74, 78, 115, 131, 133, 134, 155, 156, 157, 158, 159, 160, 161, 163, 164, 181
homework, xi, 38, 102, 103
human behavior, 167, 177
human dimensions, 167
human information processing, xiii, 165, 170

I

immersion, 172, 180
information technology, 180
interface, 22, 24, 44
internet, v, vii, ix, 22, 23, 35, 41, 42, 49, 52, 56, 61, 62, 63, 64, 65, 66, 68, 69, 70, 72, 74, 76, 96, 107, 115, 125, 128, 131, 138, 141, 149, 153
internet experience, 62, 65, 68, 69
intrinsic motivation, 6, 174, 180

L

languages, viii, 1, 31, 54, 56
laptop, 48, 90, 91, 94, 96, 149
lazy user theory, 80, 89, 90, 92, 98
leaderboard, 47
learn curricular contents, 24
learning achievement, 108, 109, 110, 112, 179
learning activity, 96, 105
learning and diagnosis, xi, 102, 103, 105, 106, 107, 112, 113
learning attitudes, 116, 118, 119
learning environment, x, 3, 9, 28, 36, 47, 65, 75, 101, 103, 113, 114, 118, 136, 160, 171, 181
learning motivation, xi, 102, 105, 108, 109, 110, 111, 112
learning outcomes, viii, 2, 42, 46, 83
learning performance, 55, 104, 107, 108, 112, 113, 115, 118, 159
learning process, viii, xii, 6, 8, 33, 37, 39, 41, 46, 52, 62, 63, 74, 109, 110, 111, 113, 122, 123, 151, 169, 175
learning styles, 47, 73, 81
learning task, 28, 141

M

mechanisms of autonomy, 41
media, 63, 73, 74, 77, 147, 153, 157, 173, 174, 177
micro story, 44
microlearning, 34
Microsoft, 7, 24, 82
m-learning, v, x, 76, 79, 80, 96, 122, 124, 126, 130
mobile communication, 129
mobile device, viii, ix, xii, 2, 7, 8, 10, 22, 23, 24, 29, 33, 34, 35, 37, 39, 44, 47, 48, 49, 50, 51, 52, 53, 55, 61, 62, 63, 64, 66,

68, 70, 72, 74, 83, 104, 113, 116, 118, 119, 122, 123, 124, 125, 126, 128, 130, 133, 135, 136, 137, 138, 139, 141, 147, 148, 149, 150,151, 154, 155, 159, 169, 170, 175, 177, 178, 183
mobile experiences, v, vii, ix, 61, 62, 63, 66, 74
mobile internet, 23, 35
mobile learning, v, vi, vii, ix, x, xi, xiii, 29, 33, 34, 35, 36, 37, 38, 50, 52, 53, 54, 55, 61, 62, 63, 64, 65, 66, 68, 74, 75, 76, 77, 78, 79, 80, 81, 82, 83, 87, 88, 89, 90, 91, 92, 96, 97, 98, 101, 102, 103, 117, 121, 122, 123, 124, 126, 130, 131, 155, 158, 160, 164, 165, 166, 167, 168, 170, 171, 172, 174, 175, 176, 179, 180, 181, 182, 183
mobile phone, 48, 64, 80, 91, 122, 129
mobile technology, vi, x, 7, 37, 65, 83, 101, 102, 103, 104, 108, 112, 113, 115, 118, 149
motivation, v, xi, xiii, 1, 2, 5, 6, 19, 22, 27, 30, 39, 44, 45, 46, 47, 52, 102, 105, 108, 109, 110, 111, 112, 115, 118, 125, 136, 141, 145, 151, 155, 159, 160, 166, 173, 174, 175, 179, 180, 182
multimedia, 127, 129, 130, 168, 169

N

network operators, 82
networking, 69, 71
next generation, 65, 181

O

online learning, 107, 113, 134, 174
operating system, 4, 24, 137

P

participants, 3, 22, 28, 36, 38, 66, 68, 81, 85, 86, 89, 92, 93, 94, 95, 96, 108, 113, 141, 143
participatory culture, 31, 39, 40
password, 11, 88
perceptions of the game, 28
personal computers, 128
playing, 3, 5, 7, 9, 11, 12, 14, 17, 19, 23, 25, 26, 27, 28, 124, 134, 173, 174, 179
portability, x, 101, 102
prior knowledge, 105, 106, 107, 109, 113, 115, 116
problem solving, 8, 114
programming, 11
psychology, 5, 30, 31, 76, 116, 156, 177, 180, 183

Q

questionnaire, ix, 22, 23, 34, 48, 50, 92, 108, 109, 110, 111, 143, 144
quiz-game, 9, 10, 23, 24, 28
quizzes, 9, 42, 47

R

reading, viii, 2, 5, 6, 14, 38, 44, 45, 49, 116
reality, ix, 34, 40, 130, 176, 177, 179, 183
resources, x, 2, 3, 28, 40, 41, 56, 62, 64, 94, 96, 101, 103, 107, 113, 128, 137, 146, 148, 150, 157, 168, 171, 172, 178
response, 42, 46, 139, 148
rewards, 5, 6, 12, 39, 46, 47, 155, 168
rubrics, 43

S

school activities, 6, 29, 50, 52

school community, 37, 41
science, vii, xi, xii, 102, 114, 115, 116, 117, 119, 121, 122, 127, 130, 167, 169, 170, 176, 178, 179, 181, 183
science education, vi, vii, xi, xii, 121, 122, 126, 127, 130, 131, 181
self-learning, 104, 109
serious games, 7, 8, 9, 28, 31, 45, 167, 168, 175, 179
servers, 11
social change, 35
social construct, 65
social context, 30
social development, 2
social interaction, 19
social learning, 27, 38, 104
social network, 23, 35, 48, 49, 129, 138, 160
social relations, 4, 123
software, xi, 24, 70, 73, 86, 102, 104, 108, 110, 112, 113, 127, 134, 137, 139, 155, 159, 170, 175
software engineering, xi, 75, 102, 104, 108, 110, 112, 113
student motivation, 136, 145

T

teachers, x, 7, 11, 28, 37, 39, 40, 41, 42, 52, 54, 63, 101, 102, 103, 104, 105, 106, 107, 108, 113, 123, 125, 127, 129, 139, 150
teaching strategies, ix, 34, 39
teaching/learning methodologies, 122
technological advances, 35, 166, 167, 170, 177
technological developments, 177
technological revolution, 38
technology, ix, x, xii, 7, 9, 26, 35, 37, 38, 39, 42, 48, 53, 61, 64, 65, 75, 76, 77, 78, 79, 81, 83, 93, 94, 95, 97, 98, 101, 102, 103, 104, 108, 112, 113, 114, 115, 116, 117, 128, 131, 133, 134, 136, 137, 138, 141, 143, 144, 145, 146, 147, 148, 149, 154, 155, 156,157, 158, 159, 160, 161, 164, 168, 173, 177, 179, 180, 181, 182, 183
technology acceptance, ix, 61, 108, 173
technology acceptance model, 108, 173
testing, 43, 105, 106, 114, 116, 117
traditional classroom, xi, 102, 103, 113, 115, 118, 127

U

universities, xii, 65, 84, 88, 133, 134, 137, 148

V

video games, 3, 4, 5, 7, 25, 28, 30, 31, 180, 182
videos, 38, 47, 49, 124, 129
visual processing, 178
vocabulary, 42, 169
vocational education, ix, 34, 39

W

web, xi, 22, 42, 86, 88, 90, 102, 105, 106, 107, 108, 109, 110, 111, 113, 114, 123, 124, 126, 142, 156
web browser, 106, 107
web sites, 124, 129
web -based clicker, 42
wireless connectivity, 153
wireless devices, 64, 157
writing, 5, 43, 44, 45